Charles Seale-Hayne Library
University of Plymouth
(01752) 588 588
LibraryandITenquiries@plymouth.ac.uk

SCALE PROBLEMS IN HYDROLOGY

WATER SCIENCE AND TECHNOLOGY LIBRARY

SCALE PROBLEMS IN HYDROLOGY

Runoff Generation and Basin Response

Edited by

V. K. GUPTA

Department of Civil Engineering, University of Mississippi

I. RODRÍGUEZ-ITURBE

*Graduate Program in Hydrology and Water Resources,
Universidad Simón Bolivar, and
Instituto Internacional de Estudios Avanzados, Caracas, Venezuela*

and

E. F. WOOD

Director, Water Resources Program, Princeton University

D. REIDEL PUBLISHING COMPANY

A MEMBER OF THE KLUWER  ACADEMIC PUBLISHERS GROUP

DORDRECHT / BOSTON / LANCASTER / TOKYO

Library of Congress Cataloging in Publication Data

Scale problems in hydrology.

(Water science and technology library)
Includes index.
1.   Watersheds—Measurement—Congresses. 2.   Runoff—Measurement—
Congresses. 3. Slopes   (Physical   geography)—Measurement—Congresses. I.
Gupta,   Vijay   K.   II.   Rodríguez-Iturbe,   Ignacio.   III.   Wood,   Eric   F.   IV.
Series.
GB980.S27   1986   551.48   86–13781
ISBN 90–277–2258–7

Published by D. Reidel Publishing Company,
P.O. Box 17, 3300 AA Dordrecht, Holland.

Sold and distributed in the U.S.A. and Canada
by Kluwer Academic Publishers,
101 Philip Drive, Assinippi Park, Norwell, MA 02061, U.S.A.

In all other countries, sold and distributed
by Kluwer Academic Publishers Group,
P.O. Box 322, 3300 AH Dordrecht, Holland.

Printed in The Netherlands

TABLE OF CONTENTS

PREFACE

A special workshop on scale problems in hydrology was held at Princeton University, Princeton, New Jersey, during October 31-November 3, 1984. This workshop was the second in a series on this general topic. The proceedings of the first workshop, held in Caracas, Venezuela, in January 1982, appeared in the *Journal of Hydrology* (Volume 65:1/3, 1983). This book contains the papers presented at the second workshop.

The scale problems in hydrology and other geophysical sciences stem from the recognition that the mathematical relationships describing a physical phenomenon are mostly scale dependent in the sense that different relationships manifest at different space-time scales. The broad scientific problem then is to identify and formulate suitable relationships at the scales of practical interest, test them experimentally and seek consistent analytical connections between these relationships and those known at other scales. For example, the current hydrologic theories of evaporation, infiltration, subsurface water transport and water sediment transport overland and in channels etc. derive mostly from laboratory experiments and therefore generally apply at "small" space-time scales. A rigorous extrapolation of these theories to large spatial and temporal basin scales, as mandated by practical considerations, appears very difficult. Consequently, analytical formulations of suitable hydrologic theories at basin wide space-time scales and their experimental verification is currently being perceived to be an exciting and challenging area of scientific research in hydrology. In order to successfully meet these challenges in the future, this series of workshops was initiated.

Even though the scale problems described above can be safely speculated to manifest themselves in most, if not all, components of the hydrologic cycles, the papers in this book mostly concern themselves with process of runoff generation and structure of the runoff hydrographs from different subbasins of a basin, including the basin itself. Typically the various papers presented in this book consider one or more aspects of spatial variability in river basins due to geology, channel network geomorphology, soil and vegetation and space-time variability in the climate via rainfall in formulating and testing relationships at large scales, governing runoff generation from hillslopes and the structure of basin hydrographs.

New theoretical advances in the descriptions of hydrologic processes at large scales require developing testable hypotheses and performing suitable experiments for testing these hypotheses. These issues pose problems whose solutions mandate that a group of scientists of some critical size from all over the world participate in both experimental and theoretical aspects of the scale problems and meet periodically in taking stock of the progress as well as identifying outstanding problems. We hope that these workshops will motivate a greater interdisciplinary participation from the international scientific community in this very important area of hydrologic science.

The four days of workshop provided an opportunity for an exciting and free exchange of ideas and long useful discussions. We thank Princeton University for its hospitality and the University of Mississippi for its administrative support. The hospitality and help extended by the Utah Water Research Laboratory at Utah State University during the final stages of preparation of this book are gratefully acknowledged. This workshop was supported by the National Science Foundation through grant CEE-84-17682 and by the Army Research Office through grant 22323-GS-CF. Sincere gratitude is expressed to these two agencies for without their support this workshop would not have been possible.

Vijay K. Gupta (University of Mississippi)
Ignacio Rodríguez-Iturbe (Caracas, Venezuela)
Eric F. Wood (Princeton University)

1

ON THE RELATIVE ROLE OF HILLSLOPE AND NETWORK GEOMETRY IN HYDROLOGIC RESPONSE

Oscar J. Mesa
Edward R. Mifflin

ABSTRACT

A simple model is developed to investigate the relative role of network geometry and hillslopes in basin response. The network geometry is quantified in a function, termed the width function, that reflects the distribution of runoff with flow distance from the outlet. The model consists of two components: the routing component of the initial distribution (width function) through the network by means of a simplified diffusion approximation; and the hillslope component. The latter is also considered in an idealized manner with the purpose of investigating the relative importance of both components. Various assumptions are tested through qualitative comparison of the model output with a recorded event.

1. INTRODUCTION

The channel network within a basin forms as a result of runoff generation and flow to the outlet. The transport of runoff to the outlet is governed by the channel network that has been formed. This interconnection suggests that the network geometry holds important clues to discovering how and why a particular network forms as well as how that particular network responds to rainfall input. This idea forms the basic thesis of this paper. Specifically, the relative influence of hillslopes and the channel network on basin hydrograph is investigated to demonstrate how the network provides qualitative physical insights into both runoff generation and its transport to the outlet. In this sense, the investigation in this paper can be termed as applicable at the basin scale in so far as it neither uses hydrodynamical formulations nor extrapolates a model from a single hillslope to the entire basin (see Gupta et al., 1986, for a further discussion of scale ideas for a river basin).

A quick survey of the classic advances in hydrology reveals clear attempts to develop theories directly at the basin scale: the time area diagram (Ross, 1921); the unit hydrograph

1

V. K. Gupta et al. (eds.), Scale Problems in Hydrology, 1–17.

(Sherman, 1932); the earlier linear models of Zoch (1934, 1936, 1937), Clark (1945), and Dooge (1959); and the geomorphic considerations of Horton (1945) and others.

Recently, Rodriguez-Iturbe and Valdés (1979) and Gupta et al. (1980), provided a firm argument in support of the identification of the instantaneous unit hydrograph (IUH) or the basin response function with the probability density function (pdf) of the basin holding time. These two terms are used interchangeably in the remainder of this paper. The importance of this identification is twofold; first, it provides the proper setting for a physically based inquiry into the density of the basin holding time, and second, it opens the idea of scales by the introduction of the random character of the holding time resulting from the interactions of many other processes at lower scales. Even though the manner of treating network geometry in these models subsequently required modifications (see Gupta and Waymire, 1983), it is interesting to observe that this re-examination of the earlier work has led to new results (see, e.g., Troutman and Karlinger, 1984, 1985, 1986; Mesa, 1982; Mifflin, 1984; Gupta et al., 1986).

In the next section we develop a simple representation for the basin holding time pdf, i.e., the basin response. This consists of a derivation of the holding time pdf of a drop of water once it enters the channel system. This is called the network response function f_n. This pdf is computed using the network geometry and a single routing function governing water transport from any point in the network to the outlet. The structure of hillslope response function is only investigated qualitatively. After a brief discussion of the runoff producing mechanism on the hillslope regions, a general representation of the hillslope response is given as a weighted sum of a fast response and a slow response. Finally the basin response is obtained as a convolution of the hillslope and the network response functions.

Section III describes the small basin in northern Mississippi used in this study to test the ideas presented in Section II. The parameters of the network and the hillslope pdf's are estimated and the network geometry is quantified. Various combinations of parameters are tested and the resulting simulated responses are compared with an observed hydrograph to obtain physical insights into the relative roles of hillslope and network responses on the basin IUH. This paper is concluded with directions for further research.

2. THE REPRESENTATION

We begin with a short presentation of a model governing water transport from any point of the network to the outlet. This presentation leads to an explanation of the width function as a compact way of representing the network geometry (see, e.g., Kirkby, 1976). We then combine these two elements to represent the network response function f_n. Next a physical discussion is given as a motivation to obtain a representation of the hillslope response function f_h. The network and the hillslope response functions are convoluted to arrive at the representation of the basin response function f_b.

2.1. Network Response

The well known Saint-Venant equation of momentum balance along with the equation of continuity are widely considered as adequate representations of water transport in an open channel (see, e.g., Dooge, 1973). However, explicit analytical solutions of these equations are difficult to obtain. So their usefulness becomes restrictive for certain analytical studies such as this one. Physically speaking, the Saint-Venant equation seems inadequate to describe transport processes in a network of natural channels in a given basin because it generally applies to transport of clear water in a straight reach of a prismatic, impervious, and non-eroding channel. For natural channels, issues become more complex since various physical features not considered in the Saint-Venant equation manifest. These include regular and irregular changes in channel geometry via meandering, changes in roughness, cross sectional areas, and direction of flow both in space and time and as a consequence in the flow itself. In addition, erosion and sediment transport mostly accompany water flow. These, along with other factors such as the interaction of a channel with the surrounding aquifer and the overbank storage during high flows, significantly influence water transport. All of these complexities manifest in a single reach of a natural channel.

In a river basin, many channel segments (or links) are interconnected to form a network. These links generally differ from one another in their geometric and hydraulic characteristics and together produce a complex picture of boundary conditions. Applying the Saint-Venant equation to water transport over a network not only becomes a formidable mathematical problem but also remains physically inadequate. A further physical complication arises because parameters in the Saint-Venant equation can only be estimated from sparse measurements and largely become "fitting parameters".

In view of the limitations alluded to above in describing movement of water in channel networks, via the Saint-Venant equation, our focus in this study is to select a function which satisfies certain general physical features of water transport. For this purpose, Gupta et al. (1980) describe the probabilistic definition of the instantaneous response function or the Greens function, $h(x,t)$, as the holding time density of a perturbation (delta function) placed at time zero at a distance x from the place of observation. From this perspective, the properties of the function h should exhibit certain general physical features of water movement. For example, the mean with respect to h should be such that the farther away the perturbation, the longer it takes to traverse the channel, and the larger the flow velocity, the smaller the mean time to traverse the flow distance. Since the variance governs the spread of the holding time density, i.e., the attenuation of the wave at the downstream end, it should increase with the flow distance and decrease with an increase in the flow velocity. The probabilistic interpretation of the Green's function h implies that two particles placed simultaneously at the same upstream distance x from a point do not necessarily arrive together at that point.

In this study, the Green's function $h(x,t)$ governing water transport in a channel is taken to be

$$h(x,t) = \frac{x}{\sqrt{2\pi}b^2t^3} \exp\left\{ -\frac{(x-at)^2}{2b^2t} \right\} \tag{1}$$

where a is the drift velocity and b^2 is the diffusion coefficient. Both a and b are positive. This function is a probability density function in t for x different from zero. It has various names in the literature such as an inverse Gaussian density, a first passage time density of a Brownian motion, a stable distribution with exponent $1/2$ (see Feller, 1971, p. 175). Some of the properties of h are

$$\text{mean} = x/a \tag{2a}$$

$$\text{variance} = b^2x/a^3 \tag{2b}$$

$$\text{skewness} = 3b/(ax)^{1/2} \tag{2c}$$

$$\text{time to peak} = \frac{x^2}{3b^2 + (9b^4 + 4a^2 x^2)^{1/2}} \tag{2d}$$

The moments of this function reflect the physical features discussed above. This function is a solution of the well known boundary value problem

$$\frac{1}{2} b^2 \frac{\partial^2 h}{\partial x^2} = \frac{\partial h}{\partial t} + a \frac{\partial h}{\partial x} \tag{3}$$

$$h(x,0) = 0 \quad \text{for} \quad x > 0$$

$$h(0,t) = \delta(t) \quad \text{for} \quad t \geq 0$$

It is known in the literature that, for large times, the diffusion equation is a linearized approximation of the Saint-Venant equation (Lighthill and Whitham, 1955). However, our earlier discussion suggests that much caution is necessary in reading too much into such connections for water flow in natural rivers and channels. For the present purposes this connection is not of much physical importance except for the observation that the parameters a and b^2 admit the following expressions via the Saint-Venant equation,

$$a = \frac{3}{2} v_o \tag{4}$$

$$b^2 = \frac{v_o^2}{gSF_o^2} (1 - F_o^2) \tag{5}$$

where $F_o = v_o/(gy_o)^{1/2}$ is the Froude number, v_o is the steady state (reference) velocity, S is the channel slope, y_o is the steady state (reference) depth, and g is the gravitational acceleration.

The response of the channel due to a time-varying input $U(t)$ at the upstream end can be expressed as the solution of Eqn. (3) by modifying the boundary condition in it (see Duff and Naylor, 1966, p. 123),

$$V(x,t) = \int_0^t U(t - s)h(x,s)ds , \tag{6}$$

where $h(x,s)$ is the Green's function given by Eqn. (1). In fact Eqns. (3) and (6) imply that

$$h(x_1 + x_2,t) = \int_0^t h(x_1,t - s)h(x_2,s)ds \tag{7}$$

To extend the response representation from a single channel to a network of channels, one may be tempted to model each and every link segment with a different Green's function h, parameterized according to the different link lengths, slopes, velocities, etc. Such an approach would be the equivalent in spirit to rainfall-runoff modeling in current use, i.e., modeling too much detail of certain processes based on sparse measurements accompanied by very little or no consideration of other equally important processes. These scale considerations lead us to explore the idea of representing the response of every point in the whole network by a single Green's function h. The response of the network to an instantaneous input of water in the channels can be expressed as

$$f_n(t) = \int_0^\infty h(x,t) W(x) dx \tag{8}$$

The weight function $W(x)$ reflects the distribution of runoff produced at a distance x in the basin. The Green's function captures the physical picture of translation and attenuation in the water movement in the network.

Technically speaking, the representation of the network response function f_n can also be understood as reflecting the randomization of the parameter x in $h(x,t)$ according to the probability density $W(x)$. The weight function $W(x)$ in Eqn. (8) can be obtained from the network geomorphology. For this, let the number of links in the network at a flow distance x from the outlet be denoted by the width function $N(x)$ (see, e.g., Kirkby, 1976). Let the total length of all the channels in the network be denoted by L_T. Then the normalized width function $(1/L_T)(N(x)$ can be viewed as the probability density of water injected randomly into the network at a distance x from the outlet. The assumption that $W(x) = (1/L_T)N(x)$ implies that the runoff generated in any local region is proportional to the stream length in that area. This simple and compact way of representing the network geometry by the width function is connected to the classical time-area diagram concept under the assumptions that the drainage density of the basin is constant everywhere and that the flow velocity is a constant.

An alternate way of expressing the weight function, $W(x)$, in Eqn. (8) comes from re-examining the above assumption that lateral inflow is proportional to the channel length. Because of inflow from the above, in each interior link the channels exhibit more meandering in the downstream reaches. However, this increase in the channel length in the downstream reaches need not be accompanied by more lateral inflow. One way to introduce this feature into the weight function is to weight the link length by the inverse of the magnitude of the link. Magnitude is used here as a measure of the total flow generated above a link. There-fore, an increase in the length of the channel link in the downstream reaches does not propor-tionately result in an increase of the lateral inflow. Weighting by the inverse of the magni-tude reflects a discounting of the runoff generated from the upstream of a link.

2.2. Hillslope Response

The role of hillslope response on the basin hydrograph is investigated in this study to obtain physical insights into the relative influence of network geometry and hillslopes on basin response.

Recent developments in hillslope hydrology identify different physical mechanisms of runoff production. For example, Horton's overland flow occurs when rainfall intensity exceeds the infiltration capacity. Subsurface flow occurs when infiltration capacity is high enough to absorb all but the rarest, most intense storms. Saturation overland flow is produced when the soil becomes saturated upwards throughout its depth and saturation reaches the ground sur-face causing rainfall on those areas to run off.

Finally, the return flow, related to variable infiltration rates, is the subsurface flow that reaches the ground surface in the zones where saturation from below has occurred. However, it is difficult to distinguish it in field situations. Because of the different mechanisms involved with different forms of runoff production, usually different flow velocities are associated with these different flows (Dunne, 1983). For example, Horton flow, typical in zones with little vegetation, is characterized by velocities on the order of 10 to 500 m/h. Subsurface flow is slow, having typical velocities on the order of 0.0001 m/h or less. Saturation overland flow and return flow at the surface travel slower than Horton flow because of the densely vegetated surfaces and gentle slopes associated with them. Velocities of these flows are in the range of 0.3 m/h to less than 100 m/h. In this study we attempt to represent these features of the hillslope component in a qualitative manner and provide some order of magnitude estimates for the hillslope response function. We do not attempt to reproduce the details of the mechanisms of water transport and runoff generation in the hillslope regions.

The probability density of the holding time T_h of a drop of water in the hillslope is taken to be distributed as the weighted sum of two densities: f_{hf} denotes the fast overland flow and f_{hs} the combination of the slower surface and ground water flow. The weights, π_f and π_s, associated with these densities, can be thought of as the probabilities with which a drop of water would take either of these two different paths to a channel from a hillslope. Note that one may allow for the sum ($\pi_f + \pi_s$) to be less than one to reflect losses. Also, the weights need not be constants, allowing for changes in antecedent conditions and similar considerations. Staying within the preliminary scope of this work, only constant weights summing to unity are considered. This decomposition of the hillslope holding time density due to two different mechanisms yields the following representation of the hillslope response function.

$$f_h(t) = \pi_f \ f_{hf} + \pi_s \ f_{hs}. \tag{9}$$

In the present study no attempt is made to characterize the densities f_{hf} and f_{hs} by physically measurable quantities.

2.3. The Basin Response

The basin IUH, f_b is taken to be the convolution of the network response function, f_n, and the hillslope response function, f_h. This is a simple consequence of the fact that the basin holding time T_b is the sum of the network holding time and the hillslope holding time T_h and the assumption that these two are independent. Therefore we can write,

$$f_b(t) = \int_0^t f_n(t-z) \ f_h(z) dz. \tag{10}$$

To summarize the developments, recall that f_n is completely specified once $W(x)$ via the network geometry and the parameters a and b^2 in the transport function $h(x,t)$ are specified. The quantification of hillslope response is more illusive. In this the two densities correspond to two different physical mechanisms of runoff production and the corresponding weights describe the proportion of water being generated by each of these two mechanisms. These developments show that even a relatively simple physical picture of basin response involves various parameters which may be difficult to measure or specify. However, the next section illustrates how these ideas lead to physical insights into the relative roles of network response and hillslope response in basin hydrographs.

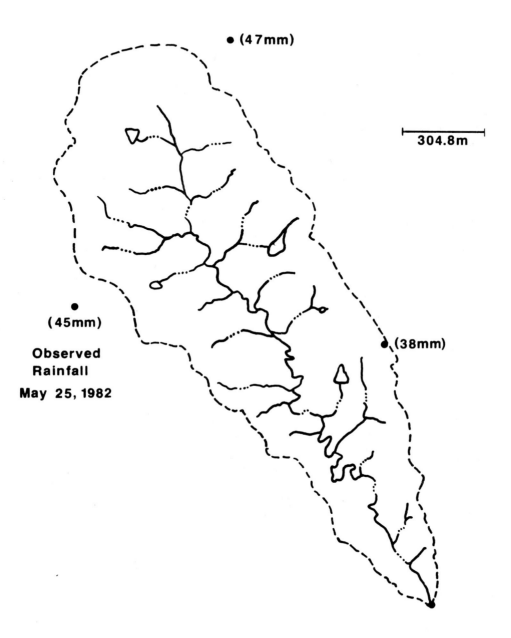

● (47mm)

304.8m

● (45mm)

Observed
Rainfall

May 25, 1982

●(38mm)

Figure 1   Drainage Network of the Goodwin Creek, Sub-basin 13, Mississippi, U.S.A.

3. MODEL TEST

Goodwin Creek is a U.S. Department of Agriculture experimental watershed. Fourteen gaging stations are located within the 21.39 km^2 basin. Each station records continuous rainfall, flow, and sediment data. There are 12 additional rain gages within the watershed boundaries.

A map of the Goodwin Creek sub-basin 13, including the location of the four closest rain gages, along with the recorded rainfall, is presented in Fig. 1. The geometric and topographic details of the channel network are given in Mifflin (1984). Examining the flow records for water years 1981 and 1982 revealed that gage No. 13 consistently showed multiple peaks. The 26 rain gages were examined in order to find an event consisting of a short duration, high intensity, uniform rainfall. A 30-minute event ranging from 29.7 to 48.8 mm of total rainfall occurred over the watershed on May 25, 1982. The four rain gages closest to sub-basin 13 had a range from 38.1 to 47.0 mm as shown in Fig. 1. To understand the relative roles of network geometry and the hillslope response in producing the multiple peak feature in the observed hydrographs, sub-basin 13 was chosen as the model basin.

Sub-basin 13 drains an area of 1.24 km^2. The basin consists of rolling hills improved as cropland and grazing land. During the 19 days prior to May 25, 1982, no runoff was recorded at the gage. During the first 20 minutes of rainfall essentially no flow occurred at the outlet. As a result, the hydrograph from this storm is assumed to reflect the features of the basin IUH for this basin. The hydrograph for this event is reproduced in Fig. 2. The time axis is in minutes from the beginning of rainfall and the flow is in m^3/s. The following assumptions about the effective rainfall were made to facilitate comparison of computed IUH's with the observed event which is the response to a 30 minute duration of rainfall. The total volume under the observed hydrograph was assumed to give the volume of effective rainfall. The effective rainfall was assumed to be concentrated as a delta function at 25 minutes from the beginning of the rainfall. This timing was suggested by the observed 20 minutes with no flow. The computed hydrographs presented later were obtained from the corresponding computed IUH by convolution with the assumed effective rainfall.

In order to apply the ideas presented in the previous section to this particular sub-basin, first, we present the construction of the width function and of the magnitude weighted width function as two possible candidates to represent the function W in Eqn. (8). Then, the parameters a and b^2 of the inverse Gaussian kernel h, given in Eqn. (1), are chosen in a qualitative way for computing the network response function via Eqn. (8). Some variations in the parameters a and b^2 are considered with the sole purpose of gaining physical understanding of the effect of the kernel on the network response function. The hillslope response is considered next. Two alternatives are considered. First, when the weight π_f in Eqn. (9) is zero, and the second when $\pi_f > 0$. These two alternatives are combined with the network IUH via Eqn. (10) to analyze features of the computed basin IUH and compare them with the observed hydrograph.

The width function of a given network is constructed as follows. First the links are labeled in consecutive order from 1 to 2 m-1, where m is the magnitude of the basin. Let the flow distances from the downstream and upstream edges of the kth link to the outlet be denoted by α_k and β_k respectively. Define for any two numbers α and β, $\alpha < \beta$, the indicator function $\mu(x;\alpha,\beta)$ to be

$$\mu(x;\alpha,\beta) = 1 \quad \text{if} \quad x \in [\alpha,\beta] \tag{11}$$

$$= 0 \quad \text{otherwise.}$$

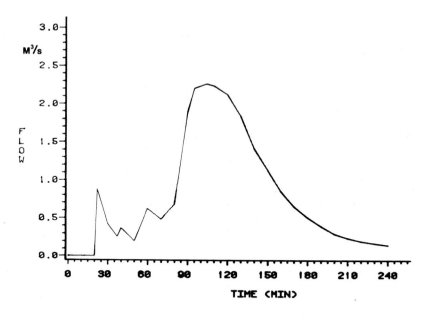

Figure 2   Observed Discharge Hydrograph. Goodwin Creek, Mississippi,
                    Sub-basin 13.   May 25, 1982.

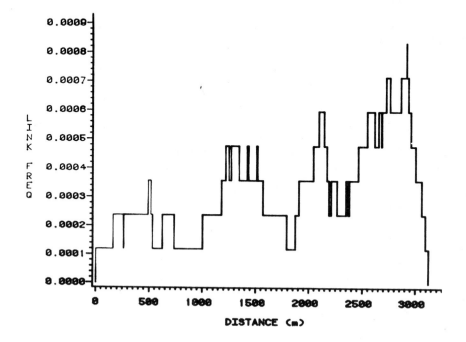

Figure 3   Normalized width function for the Sub-basin 13 of Goodwin Creek.

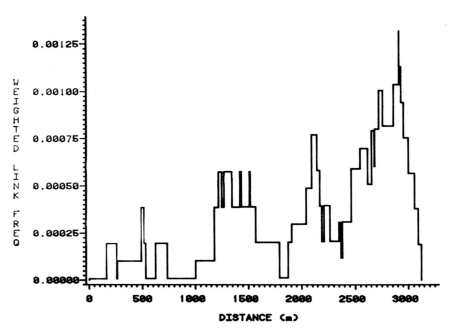

Figure 4   Magnitude Weighted Width Function for the Sub-basin 13 of Goodwin Creek.

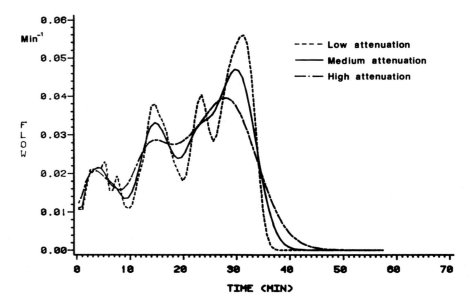

Figure 5   Network Response Function for Three Sets of Attenuation Parameters

The width function $N(x)$ can now be expressed as,

$$N(x) = \sum_{k=1}^{2m-1} \mu(x;\alpha_k,\beta_k). \tag{12}$$

Our first candidate for W is the normalized width function given by

$$W(x) = \frac{1}{L_T} N(x) \tag{13}$$

where L_T is the total channel length. $W(x)$ for sub-basin 13 is presented in Figure 3.

As discussed earlier, if instead of considering the runoff to be proportional to channel length as in Eqn. (13), we compensate for the upstream inputs by weighting inversely by the magnitude, a second representation of W is obtained which can be expressed as,

$$W^*(x) = \frac{1}{D} \sum_{k=1}^{2m-1} \frac{1}{m_k} \mu(x,\alpha_K,\beta_K) \tag{14}$$

where D is chosen so that the area under W^* is one and m_k is the magnitude of the $k-th$ link. Figure 4 shows the inverse magnitude weighted width function for the sub-basin under consideration. Comparing W with W^*, the effect of giving less weight to the main channel and other high magnitude links can be easily seen. Some observations which are similar in spirit to computing the magnitude weighted width function are given in Rogers (1972).

Now for computing the two parameters a and b^2 in h, Eqns. (4) and (5) are used as a guide. For example, a typical value of 1 m/s is considered reasonable for v_o. In order to compute the slope, S in Eqn. (5), the topographic map is used to find the average link slope to be around 0.03 and an average slope to the outlet from any point in the basin to be about 0.01. During a field trip the flow in the channels was generally observed to be tranquil. This led us to specify a typical value of the Froude number F_o to be 0.6. These values do not reflect the hydraulic characteristics of any single reach of a channel. Rather they represent an average for the network. Since we do not as yet have a physical basis of specifying h and the parameters therein, three sets of parameters are considered to investigate their relative influence on the network response f_n. The velocity v_o is fixed at 1 m/s throughout. The other two parameters F_o and S in Eqn.(5) control the attenuation. The three sets of these parameters considered here are labeled low, medium, and high. In the medium set, S is 0.01 and F_o is 0.6, as explained above. The other two extremes are $S = 0.008$ and $F_o = 0.4$ for high attenuation and $S = 0.03$ and $F_o = 0.8$ for low attenuation.

Figure 5 shows the simulated network response for the low, medium and high attenuation parameter sets when the initial function is given by Eqn. (13). Figure 6 shows the network response for the case of medium attenuation parameters when the initial function is given by Eqn. (14). The effect of the attenuation in smoothing the network geometry displayed in Figs. 3 and 4 is evident as most of the fluctuation in these functions get smeared. This is particularly true for fluctuations in W at large flow distances. Some details of the function W at smaller distances are retained in the network response but these details progressively undergo smearing with increasing attenuation. Also, the time base of the network response is about one fourth of that of the observed hydrograph. This suggests that the hillslope response produces the recession in the observed hydrograph at the basin outlet. This effect is more pronounced here because of the small size of the basin.

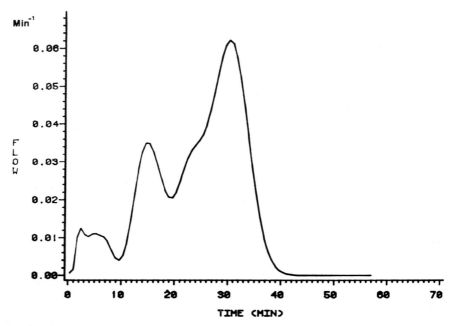

Figure 6   Network Response Function for Medium Attenuation Parameters and Magnitude Weighted Width Function.

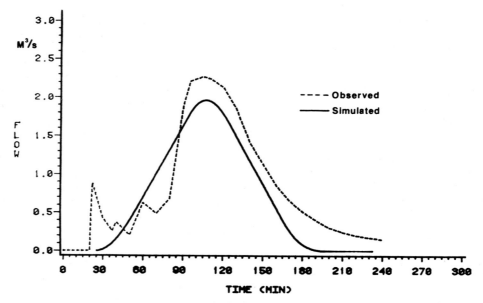

Figure 7   Observed and Simulated Basin Hydrographs; the simple hillslope response was used in the computed hydrograph.

The hillslope response will now be introduced to facilitate an assessment of the relative importance of the network and the hillslopes on the observed hydrograph. First, a comparison of the observed hydrograph with the network response function computed above provides an estimate of the difference between the means of the two hydrographs of about 70 minutes. This difference is hypothesized here to be caused by the delay coming from the hillslopes. Now the average hillslope flow distance of 80 m is obtained by taking one-half the inverse of the drainage density, an idea that goes back to Horton (1932). The observed mean difference of 70 minutes when combined with an average flow distance of 80 m, gives a characteristic velocity on the order of 0.02 m/s. This velocity is well in the range reported in the literature as discussed in the last section. A simple isosceles triangle with a mean of 70 minutes is first considered as representing the hillslope response. The hillslope response is convoluted with the network response function corresponding to the medium attenuation parameters to give the basin IUH as shown in Fig. 7. This illustrates how the base as well as time to peak of the computed hydrograph compares well with those of the observed hydrograph. It can be seen from Fig. 7 that although the recession of the basin response function agrees better with the observed hydrograph, it does not show the multiple peaks at the beginning that are present in the observed hydrograph.

The absence of the multiple peaks in the basin response computed above led us to test the hypothesis that hillslope response consists of a combination of a fast response and a slow response. This fast response can be interpreted physically as either Hortonian overland flow or saturation overland flow near the channels. For that reason the average length for the fast response is reduced from 80 m to 20 m and a small weight of 0.10 is assigned to π_f in Eqn. (9). A velocity of the order of 0.25 m/s is taken as representative of the fast response. This gives the mean of the pdf denoting the fast response in eq. 9 to be 1.3 min. The form of the pdf is taken simply to be an isosceles triangle. The slow hillslope response can be attributed to either saturation overland flow and/or subsurface flow, and therefore a delay time of 20 minutes for the response to start is reasonable. The form of the pdf for the slow response was kept as an isosceles triangle with the same mean as before (70 min) and the shift of 20 minutes from the origin as noted above. The combined hillslope response is represented in Fig. 8. The convolution of this hillslope response function with the network response function, corresponding to the medium attenuation parameters, is represented in Fig. 9. Interestingly, the introduction of the fast response in the hillslopes preserves the multiple peaks at the beginning of the computed basin response, just as they manifest in the observed hydrograph. The assumed 20 minute delay was shown to be essential to reproduce these multiple peaks. A run, not presented here, with the rest of the parameters equal, but without the delay, did not reproduce the multiple peaks.

The above study supports the observation that the hillslope alone is not responsible for the noise at the beginning of the observed hydrograph. The multiple peaks are present in the network response function (see Fig. 5). If the hillslope response is smooth, the convolution of the network and hillslope response smears out those peaks. The fast part of the hillslope response acts almost like a delta function on the network response which forms the beginning of the basin IUH.

4. CONCLUSIONS

The following conclusions can be drawn from this study.

(1) The channel network geometry significantly influences the basin hydrograph even in a "small" basin such as the one investigated here. The width function provides a good first step in quantifying the influence of the basin geometry on the hydrograph.

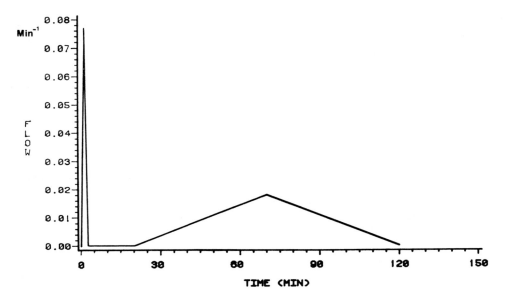

Figure 8   Combined Hillslope Response Function

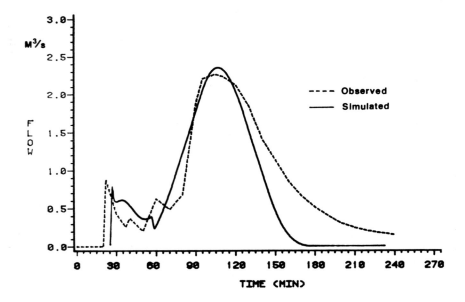

Figure 9   Observed and Simulated Basin Hydrographs; the combined hillslope response
                     function was used in the computed hydrograph.

(2)   Hillslope response consists of a combination of a fast component and a slow component. In our case study only a small fraction of the total runoff appears via the fast component while the bulk of the runoff appears in the network via a slow component.

(3)   The recession function of the hydrograph for the entire basin is controlled by the slow component of hillslope response, at least in a small basin.

(4)   The initial part of the hydrograph is controlled by the fast component of the hillslope response.

(5)   It is feasible to model the basin hydrograph directly at the basin scale. In this construction the channel network via its link structure plays the most important role. The typical physical parameters that appear in this study are the width function, link magnitude, drainage density, mean link slope and mean flow velocity for the network, and the fast slow hillslope response component weights and velocities.

The need for further research at the basin scale is clear. The results obtained in this study are encouraging enough to warrant a more detailed analysis of our formulation. For instance, we have, in some sense, replaced the basin by an equivalent channel. The desire to average the flow dynamics by envisioning this equivalent channel may not be as ill advised as one might think. Even if two tributaries at an equal (time) distance from the outlet have substantially different hydraulic characteristics, water flowing from their respective outlets shares the same path through the main channel. Therefore, it seems reasonable to envision that the common holding times on this shared path will tend to minimize any differences present in the non-shared paths. Another consideration in averaging the dynamics is the complimentary relationship between flow and channel formation. Can we identify some basin scale regularities in a network that would justify this averaging argument? Since the driving force in water transport is gravity, perhaps considering altitude relationships in the network geometry is the best place to start such an inquiry (see, e.g., Gupta et al., 1986).

The weight function $W(x)$ is perhaps the most interesting part of our formulation. One might use it to consider the problem of basin similarity. For example, if two basins have the same width function then what similarities and differences would they reflect in their respective hydrographs? Notice that the basin width function does not uniquely determine the network geometry. In other words, two geometrically different networks may possess the same width function; also see Kirkby (1976).

An important question that needs further consideration is the relative importance of the network and hillslopes on basin response. It appears that in larger basins the network component of the response may dominate the basin response. Moreover, since the attenuation due to channel routing is proportional to the distance from the outlet, one expects that the irregularities in the width function may become even less important in larger basins. Another aspect that needs further consideration is the quantification of the volume of runoff. The influence of the space distribution of soil moisture previous to the storm, as well as the space distribution of rainfall, vegetation cover and evaporation should be considered from a basin scale point of view. This particular problem was not considered in this paper.

In conclusion we wish to emphasize that the importance of network geometry is not confined only to routing, because we think that it also holds important clues to runoff generation on a basin. To investigate each hillslope separately for modeling the basin response would lead to tracking numerous parameters which would eventually be averaged out in the basin response. Therefore appropriate quantification of the channel network geometry is the right place to look for investigating the response of a basin with an identifiable channel network.

ACKNOWLEDGMENTS

This research was supported by Grant 21078-GS from the Army Research Office. Gratitude is expressed to Professor V.K. Gupta for his guidance during the course of this study and the USDA Sedimentation Laboratory, Oxford, Mississippi, for providing data for the Goodwin Creek sub-basin.

REFERENCES

Clark, C.O., 1945. Storage and the Unit Hydrograph, *Am. Soc. of Civ. Engrs. Trans.*, Vol. 110, pp. 1419-1446.

Dooge, J.C.I., 1959. A General Theory of the Unit Hydrograph, *J. Geophys. Res.*, Vol. 64, No. 2, pp. 241-256.

Dooge, J.C.I., 1973. Linear Theory of Hydrologic Systems, *Tech. Bul. 1468*, U.S. Department of Agriculture, Washington, DC.

Duff, G.F.D., and D. Naylor, 1966. *Differential Equations of Applied Mathematics*, John Wiley, New York.

Dunne, T., 1983. Relation of Field Studies and Modeling in the Prediction of Storm Runoff, In: *Scale Problems in Hydrology*, (eds.) I. Rodríguez-Iturbe and V.K. Gupta, *J. Hydrol.*, Vol. 65.

Feller, W., 1971. *An Introduction to Probability Theory and Its Applications, II*, John Wiley, New York.

Gupta, V.K., and E.C. Waymire, 1983. On the Formulation of an Analytical Approach to Hydrologic Response and Similarity at the Basin Scale, *J. Hydrol.*, Vol. 65, pp. 95-123.

Gupta, V.K., E. Waymire, and I. Rodríguez-Iturbe, 1986. On Scales, Gravity and Network Structure in Basin Runoff, Chapter 8, in: *Scale Problems in Hydrology* (ed.) I. Rodríguez-Iturbe, V.K. Gupta, E.F. Wood, Reidel Publishing.

Gupta, V.K., E. Waymire, and C.T. Wang, 1980. A Representation of an Instantaneous Unit Hydrograph from Geomorphology, *Water Resources Research*, Vol. 16, No. 5, pp. 855-862.

Horton, R.E., 1932. Drainage Basin Characteristics, *EOS Trans., AGU*, No. 13, pp. 350-361.

Horton, R.E., 1945. Erosional Development of Streams and Their Drainage Basins: Hydrophysical Approach to Quantitative Morphology, *Bul. Geol. Soc. Amer.*, Vol. 56, pp. 275-370.

Kirkby, M.J., 1976. Tests of Random Network Model, and its Application to Basin Hydrology, *Earth. Surf. Proc.*, Vol. 1, pp. 197-212.

Lighthill, M.J., and G.B. Whitham, 1955. On Kinematic Waves: I. Flood Movement in Long Rivers, *Proc. Roy. Soc. A*, Vol. 229, pp. 281-316.

Mesa, O., 1982. On an Analytical Approach for Coupling Hydrologic Response and Geomorphology, unpublished M.S. thesis, Department of Civil Engineering, University of Mississippi.

Mifflin, E.R., 1984. On the Role of Network Geometry in Basin Response, M.S. Thesis, University of Mississippi.

Rogers, W.R., 1972. New Concepts in Hydrograph Analysis, *Water Resources Research*, Vol. 8, No. 4, pp. 973-981.

Rodríguez-Iturbe, I., and J.B. Valdés, 1979. The Geomorphic Structure of Hydrologic Response, *Water Resources Research*, Vol. 15, No. 6, pp. 1409-1420.

Ross, C.N., 1921. Calibration of Flood Discharge by the Use of a Time-contour Plan, *Trans. Inst. of Engrs.*, Vol. II: 85, Australia.

Sherman, L.K., 1932. Streamflow from Rainfall from the Unit Hydrograph Method, *Eng. News Record*, Vol. 103, pp. 501-505.

Troutman, B.M., and M.R. Karlinger, 1984. On the Expected Width Function for Topographically Random Channel Networks, *J. of Appl. Prob.*, Vol. 22, pp. 836-849.

Troutman, B.M., and M.R. Karlinger, 1985. Unit Hydrograph Approximations Assuming Linear Flow Through Topologically Random Channel Networks, *Water Resources Research*, Vol. 21, No. 5, pp. 743-754.

Troutman, B.M., and M.R. Karlinger, 1986. Averaging Properties of Channel Networks Using Methods in Stochastic Branching Theory, (this issue).

Zoch, R.T., 1934. On the Relation Between Rainfall and Streamflow, *Monthly Weather Rev.*, Vol. 62, pp. 315-322.

Zoch, R.T., 1936. On the Relation Between Rainfall and Streamflow, *Monthly Weather Rev.*, Vol. 64, pp. 105-121.

Zoch, R.T., 1937. On the Relation Between Rainfall and Streamflow, *Monthly Weather Rev.*, Vol. 65, pp. 135-147.

2

NONLINEARITY AND TIME-VARIANCE OF THE HYDROLOGIC RESPONSE OF A SMALL MOUNTAIN CREEK

Elpidio Caroni
Renzo Rosso
Franco Siccardi

ABSTRACT

This paper stresses the dynamic nature of the hydrologic response of a small alpine catchment. The time-scale of the catchment response to an input of rainfall excess has been found to depend mainly on the peak flow rate of the surface runoff. A nonlinear model, providing for variable time-scale or lag-time, should therefore be parameterized in terms of geomorphologic and hydraulic characteristics of the catchment in order to represent the rainfall-runoff transformation. Moreover, nonlinearity is found to decrease with increasing storm intensity. The linear assumption can thus represent quite satisfactorily this transformation in order to perform major flood analyses also for "small" catchments, i.e., at the "elementary" basin scale. Reliable estimates of the bankful discharge are required for the purpose of estimating the average time-space streamflow velocity in this case.

1. INTRODUCTION

The problem of modelling the response of a catchment to storm rainfall can be approached through empirical hydrograph analysis, system theory and geomorphoclimatic, or analytical, theory at the basin scale.

The latter approach emanates from the practical purpose of relating the hydrologic response of a catchment to its physical characteristics, but it can also provide a deeper insight of the formation of storm runoff in comparison with the one provided by the empirical and system approaches (Rodriguéz-Iturbe and Valdés, 1979; Gupta et al., 1980). Moreover, it may indicate a route towards a unified theory of catchment response, since it can perspectively set up a theoretical background of some empirical and system achievements, thus improving the capability of their application in the hydrological practice.

V. K. Gupta et al. (eds.), Scale Problems in Hydrology, 19–37.
© *1986 by D. Reidel Publishing Company.*

For example, the two-parameter gamma distribution is a well-known analytical function providing satisfactory flexibility for the purpose of modelling the IUH (instantaneous unit hydrograph). This function was first introduced on the basis of the cascading reservoir analogue (Nash, 1960), but its capability of reproducing the GIUH (geomorphological instantaneous unit hydrograph) model derived through an analytical approach by Rodriguéz-Iturbe and Valdés (1979) has been recently proven (Rosso, 1984). Accordingly, its shape parameter can lump the effects of the physical laws determining the structure of the stream network, and its scale parameter can lump water transport dynamics at the basin scale.

A problem involved in the analysis of the hydrologic response of a catchment is concerned with the superposition and proportionality properties of the runoff hydrograph and with its constant time baselength for a given storm duration. The results obtained by assuming these empirical postulates are often in satisfactory agreement with observed data, due to the integrated effect of all catchment characteristics at the basin scale. Nevertheless, this is not a general rule, because the properties of both the ordinates and the time abscissa of the runoff hydrograph may not be confirmed by observed data. This is due to the fact that most physical processes determining the runoff hydrograph can neither be described in terms of these postulates, since their internal description is based on the equations of fluid mechanics, nor by their integrated effects at the basin scale.

In terms of the system theory, the problem is to analyze the properties of the function relating the storm runoff at the outlet of a catchment to the rainfall excess over it (Diskin, 1982). The traditional approach to this problem is generally carried out by analyzing the capability of the linear assumption to represent the rainfall-runoff relationship for a given storm. Then, the capability of the linear kernel to reproduce different storms is investigated in order to detect nonlinearity effects. Finally, its capability to reproduce the same output from given similar inputs is investigated to detect time-variance effects. A simple approach was introduced by Diskin (1973) in terms of the lag-time of the IUH.

This approach has been improved by Wang et al., (1981) on the basis of the analytical approach. A measure of nonlinearity is introduced through the dependence of the mean holding time on the rainfall intensity. Nonlinearity is found to decrease as the basin size increases, a result in agreement both with hydrologic experience and with the heuristic observation that the integrated effect of all catchment characteristics increases with basin size. A similar approach has been developed by Rodriguéz-Iturbe et al., (1982) in terms of the geomorphoclimatic theory, which allows to determine the dependence of the hydrologic response on storm intensity relying on the kinematic wave assumption associated with catchment geomorphology. The general achievement of the analytical approach has been to relate the linearity and time-variance problem to water transport dynamics at the basin scale.

The present paper deals with the problem of detecting nonlinearity and time-variance at the basin scale. Though the basin scale ranges from few hectars to several square kilometers in the hydrological practice, this scale is identified by the external descriptions used to represent the system, the operation of which is inferred from the input and output observations (Dooge, 1979). The analysis has been performed using field data from a small cátchment, where the variability in space of rainfall patterns, soil properties and vegetation cover would not appreciably affect the rainfall-runoff process. The size of this drainage basin can therefore be assumed to be representative of the smallest scale in space which needs an external approach in order to analyze the formation of surface runoff. Moreover, the integrating effects of variability and complexity do not dominate at this scale. The **elementary** basin scale, as defined by approximately uniform conditions of climate, soil and vegetation, should be suitable for measuring nonlinearity and time-variance effects due to water transport dynamics throughout a spatially homogeneous boundary.

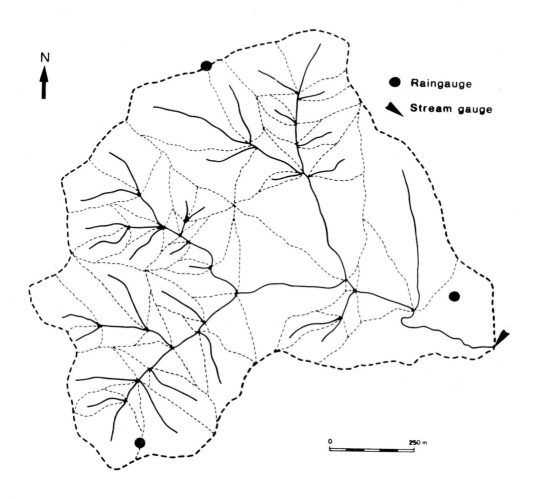

Figure 1   Map of Rio della Gallina displaying gauge locations and stream ordering.

A very simple approach has been adopted for this purpose. The two-parameter gamma model of the hydrologic response has been utilized, providing for variable time-scale of the IUH, as related to water transport dynamics. Records from field observations obtained with a satisfactory degree of accuracy, as related to the investigated scale, have been processed to analyze the catchment response to storm events.

2. FIELD AREA

2.1 General Features

From 1980, the CNR IRPI Turin Institute has carried out experimental investigations on a small creek, Rio della Gallina, located in the Prealpine Mountains area each of Biella (Italy).

This creek is tributary to Marchiazza stream, the drainage basin of which, located in the Po river valley, is under investigation according to a long-term research program on representative catchments.

Instrumentation of the creek has been carried out in order to investigate hydrological, erosional and sediment transport processes at the elementary basin scale (Caroni and Tropeano, 1981; Anselmo et al., 1982). Its drainage basin is 1.09 sq km in area, a size which is representative of the surface runoff and related processes at that scale.

The maximum relief of Rio della Gallina catchment is 196 m and the elevation at the outlet site is 326 m above mean sea level. Overland scope is 49 per cent on average. Vegetation, mainly wood with copse broad-leaved trees and minor conifers, covers the entire catchment area. There are not agricultural sites, houses and roads within this drainage basin, which can be assumed to be representative of forested mountain catchments in the Alps.

2.2 Catchment Geomorphology

Stream network analysis has been performed from aerial photographs (scale 1:5000). According to the Strahler's ordering scheme, the basin was obtained as fourth order. Figure 1 shows a map of the investigated catchment displaying the ordered structure of the stream network and contributing areas. The catchment characteristics inclusive of Horton order ratios, which have been determined by performing the graphical procedure reported in Figure 2, are shown in Table 1.

2.3 Instrumentation and Data Recording

A sharp-crested weir, double-V shaped, has been utilized to provide the measurement of the discharge at the outlet of the catchment. This prototype has been calibrated on the basis of laboratory tests by means of hydraulic model at the Hydraulic Institute of the Genoa University. This device has shown to provide satisfactory accuracy for been the entire range required for discharge measurement, varying from values less than 1 l/s to values greater than 10 cum/s.

Precipitation is measured by three raingauges, the location of which forms a triangular network with approximately equidistant gauging sites inside the catchment area, as shown in the map reported in Figure 1. The gauging devices are tilting-bucket type providing for digital data acquisition and storage.

All recording instruments are synchronized by a quartz equipment. The time sampling is presently provided to readings at 5 minute intervals. Data are made available as continuous records on digital cassette-tape in the form of 5 minutes stage readings and 5 minutes

TABLE 1. Catchment Characteristics

Drainage area, A(sqkm)	1.09
Order, (-)	4
Average (°) 1st order stream length, $L_1(m)$	151.6
Mean (§) 1st order stream length, $L_1(m)$	93.3
Mainstream length, $L\ (m)$	1570
Overland slope, OLS (%)	49
Mainstream slope, ES (%)	4.8
Bifurcation ratio, R_b (-)	3.04
Stream length ratio, R_L (-)	2.03
Stream area ratio, R_a (-)	3.92

(°) measured;
(§) extrapolated.

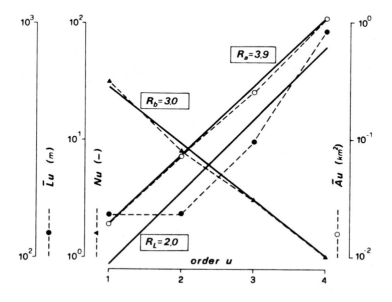

Figure 2   Graphical fitting of Horton order ratios for Rio della Gallina.

TABLE 2. Rainfall and Runoff Characteristics of Selected Flood Events

Event		R (mm)	Q_p (m^3/s)	Φ (-)	r (mm)	t_r (min)	i_r (mm/hr)	$i_r^{(5)}$ (mm/hr)	q_p (m^3/s)	q_b (m^3/s)
1.	Apr 25	87.1	1.62	.27	23.56	550	2.57	16.8	1.15	0.142
2.	May 03	40.7	0.56	.21	8.38	205	2.45	7.7	0.50	0.014
3.	Jun 28	48.7	0.68	.09	4.15	25	9.96	20.8	0.65	0.003
4.	Jul 10	46.5	2.62	.19	8.96	30	17.92	38.8	2.59	0.008
5.	Jul 22	27.7	0.83	.19	5.14	50	6.17	19.8	0.76	0.020
6.	Aug 10	66.6	8.83	.23	15.48	20	46.44	74.6	8.81	0.001
7.	Sep 10	44.2	2.02	.27	11.81	85	8.34	19.1	1.93	0.035
8.	Sep 21	73.8	2.02	.24	17.42	100	10.45	26.9	1.82	0.008
9.	Sep 24	106.2	21.60	.60	64.20	60	64.20	94.4	21.50	0.074
10.	Sep 26	26.9	1.54	.18	4.90	15	19.60	38.0	1.40	0.076
11.	Oct 26	36.7	0.57	.30	11.18	385	1.74	4.7	0.50	0.008
12.	Dec 18	15.1	0.57	.50	7.58	155	2.93	5.6	0.53	0.010

cumulative point rainfall readings. Stage readings are converted into discharge values by means of the rating curve, which has been calibrated, as reported above, to give the stage-discharge relationship for the weir.

3. ANALYSIS

3.1 Data Assembly

Twelve storm runoff events have been selected from 1981 records according to the highest observed values of the peak discharge. Since the hydrograph analysis has been found to be quite unsensitive to the method of baseflow separation, the straight line method has been utilized owing to its simplicity. The same method has been utilized to determine the rainfall excess from the observed hyetograph. In Table 2 some figures of the storm runoff events are shown whereas the meaning of all symbols used is given in Appendix A.

3.2 Shape and Time-Scale of the Hydrologic Response

The Nash model of the IUH has been utilized following the parametrization introduced by Rosso (1984), relying on the geomorphologic approach by Rodriguéz-Iturbe and Valdés (1979). Accordingly, the two-parameter gamma pdf

$$u(t) = [1/k\,\Gamma(n)]\,(t/k)^{n-1}\exp(-t/k), \quad t \geq 0 \tag{1}$$

can be parametrized in terms of Horton order ratios, of stream and catchment scales and of streamflow velocity

$$n = 3.29(R_b/R_a)^{0.78}R_L^{0.07} \tag{2}$$

$$k = 0.70[R_a/(R_b\,R_L)]^{0.48}L/v \tag{3}$$

in order to give the impulse response function of the catchment under analysis to a unit instantaneous rainfall excess over it. Such an approach provides the shape parameter, n, of the IUH, $u(t)$, to be a constant for a given catchment which can be unequivocally determined

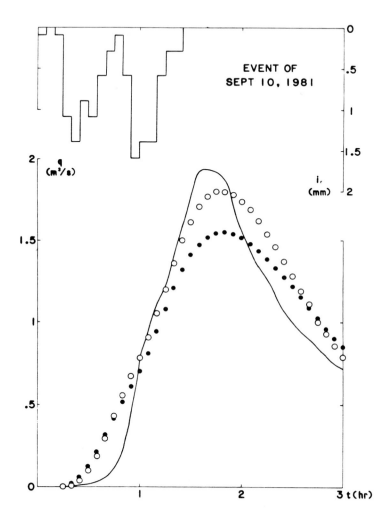

Figure 3   Comparison between observed (continuous line) and two
simulated hydrographs by the Nash model estimated by the
methods of moments (points) and least squares (circles) for
event no. 7 (10 Sep 1981).

from catchment geomorphology, namely its Horton order ratios. This property has been utilized to analyze the hydrologic response of the investigated catchment on the basis of its
time-scale k, as it results from the calibrations performed from the data.

3.3 Nonlinearity and Time-Variance as Reflected by the Time Scale of the Hydrologic Response

The experimental records have been processed in order to analyze the actual catchment
response to effective storm rainfall events. Three assumptions have been examined, namely

(i)   the rainfall-runoff process can be represented by a linear time-invariant lumped system
       (Nash, 1960; Dooge, 1973);

(ii)   the rainfall-runoff process can be represented by a linear time-varying lumped system,
       providing for variable time-scale of the hydrologic response from storm to storm, as
       related to seasonality and/or antecedent moisture conditions, and also during the storm,
       but independently from its characteristics (Chiu and Bittler, 1969; Mandeville and
       O'Donnell, 1973);

(iii) the rainfall-runoff process can be represented by a non-linear time-invariant lumped sys-
       tem providing for variable time-scale of the hydrologic response as related to the actual
       storage conditions of the catchment (Reed et al., 1975), to the actual streamflow velo-
       city throughout the stream network (Rodriguz-Iturbe et al., 1982) or to the actual inten-
       sity of the rainfall input (Wang et al., 1981).

A linear approach has been adopted for the purpose of analyzing the time scale, k, of the
response function, the shape of which has been established on the basis of its geomorphologic
estimate computed thorough equation (2), resulting n= 2.84 for Rio della Gallina. The
experimental values taken from the time-scale or from the lag-time, being

$$k = t_L / n \qquad (4)$$

somewhat reflect the accuracy of the above assumptions to represent the rainfall-runoff process, namely

(i)   k is approximately constant;

(ii)   k is variable depending on seasonality and/or antecedent soil moisture conditions, but it
       does not depend on storm and runoff characteristics;

(iii) k is variable depending on both storm and runoff characteristics.

3.4 Estimation of the Time-Scale of the Hydrologic Response

Table 3 shows the results of the analyses performed to fit the appropriate time-scale of the
IUH. First, the lag-time has been computed from the first order moments of the rainfall
excess hyetograph and the direct runoff hydrograph. The value of the time-scale, k, has been
then determined through Eqn. (4) and the average (time-space) streamflow velocity has been
computed by inverting Eqn. (3). It would be observed that this method is not able to provide
reliable results in all cases; in fact, the simulated hydrographs were found to underestimate
the observed peak discharge for some examined events, especially for the heaviest ones.

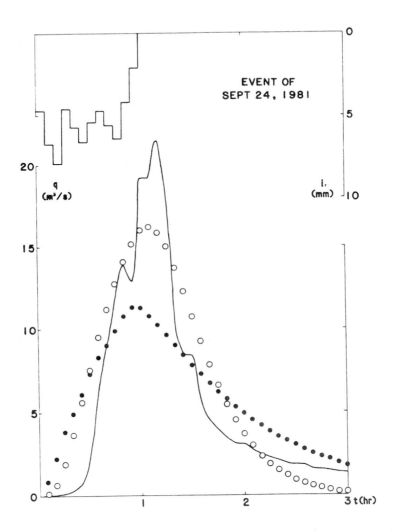

Figure 4   Comparison between observed (continuous line) and two
simulated hydrographs by the Nash model estimated by
the methods of moments (points) and least squares
(circles) for event no. 9 (24 Sep 1981).

Accordingly, a better procedure has been introduced by means of the least squares method, the analytical developments of which are reported in Appendix B. This method has been shown to improve the performance of the Nash model for the investigated catchment as shown in Figures 3 and 4.

According to Nash and Sutcliffe (1970), the coefficient of efficiency has been computed by:

$$E = 1 - \sum_j (q_j - \hat{q}_j)^2 / \sum_j (q_j - \overline{q})^2 \tag{5}$$

where $\overline{q}$ is the mean observed flow rate, q_j and $\hat{q}_j$ are the observed and simulated flow rates respectively. The resulting values are reported in Table 3, indicating such improvements.

TABLE 3. Time-Scale Parameters at the Hydrologic Response

Event	Moment	Estimates			Least Square		Estimates	
	t_L	k	v	E	t_L	k	v	E
	(min)	(min)	(m/s)	(%)	(min)	(min)	(m/s)	(%)
Apr 25	439	155	.10	1.5	153	54	.27	44.0
May 03	258	91	.16	90.8	264	93	.16	91.1
June 28	107	38	.39	73.5	95	34	.44	75.7
Jul 10	95	34	.43	64.1	73	26	.57	74.7
Jul 22	149	53	.28	32.3	239	84	.18	37.9
Aug 10	63	22	.67	60.8	51	18	.82	66.6
Sep 10	97	34	.43	90.8	85	30	.49	94.2
Sep 21	94	33	.44	83.6	90	32	.46	83.9
Sep 24	65	23	.64	72.3	44	16	.95	88.4
Sep 26	67	23	.63	88.4	63	22	.66	88.9
Oct 26	314	110	.13	89.6	297	104	.14	90.1
Dec 18	218	77	.20	88.9	204	72	.20	89.8

Both the moments and least squares estimates have been analyzed for the purpose of examining also possible effects of the method used for model calibration.

3.5 Results and Analysis

Significant variations of the time-scale of the hydrologic response are displayed by the fitted values which are reported in Table 3.

Consequently, the first assumption would provide only a rough approximation to a suitable representation of the rainfall- runoff process.

Nevertheless, the accuracy of such an assumption seems to increase with the intensity of the storm, as it can be observed in Figures 5,6 and 7.

Therefore, this should be further investigated on the basis of a larger sample of flood events in order to assess its likelihood within the range of major floods analysis.

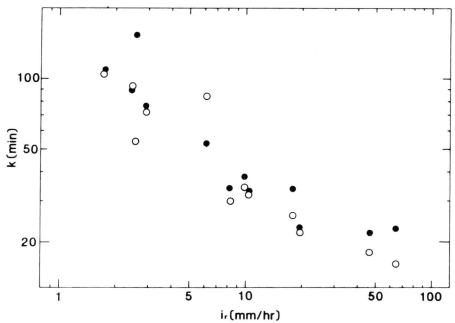

Figure 5   Time-scale, k, of the hydrologic response against average
storm intensity of the rainfall excess: moments (points) and
least squares (circles) estimates.

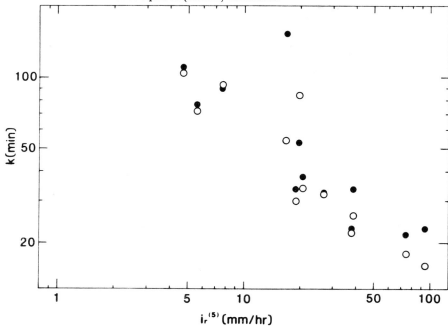

Figure 6   Time-scale, k, of the hydrologic response against
maximum (5-min duration) intensity of the rainfall excess:
moments (points) and least squares (circles) estimates.

As shown in Figure 8, seasonal variations can not be detected, at least on the basis of the investigated storms. However, the small size of the sample does not allow to assess this result, because these effects could be hidden by other factors, such as storm intensity. The time-scale of the hydrologic response seems to be also quite uncorrelated with the antecedent soil moisture conditions, as they are roughly described in terms of the initial baseflow, q_b. This latter result does not agree with the dependence of the IUH on the antecedent moisture conditions reported by, among others, Rosso and Tazioli (1979) and Pilgrim (1983) on the basis of field observations.

A general result concerning assumption (ii) is given by the undetected sensitivity of the time-scale of the catchment response with respect to any time-varying characterization independent of the storm and runoff characteristics. For example, events 2, 11, and 12 are quite similar both in the rainfall excess intensity and in the peak discharge; the same can be observed for events 7 and 8.

The dependence of the time-scale, k, of the IUH on the characteristics of the storm has been investigated on the basis of four indexes describing the intensity of the rainfall excess hyetograph, namely its average value and the maxima with duration 1 hour, 5 minutes and equal to k respectively. These figures are reported in Table 4. While the maximum 1-hr rainfall intensity has been found quite uncorrelated with the estimated values of k, the correlations displayed by the relations of the average and of the peak values with the time-scale of the IUH are not negligible, as it can be observed in Figures 5 and 6. A power function of the type suggested by Wang et al., (1981) seems to provide an appropriate functional relationship between time-scale and storm intensity, even though the variability of k estimates seems to decrease with increasing storm intensity, thus allowing to infer a limiting expression for k. The relationship between the time-scale, k, and the k-duration rainfall intensity was found to display a similar behavior, without improving the variance which can be explained through such a relationship.

TABLE 4. Indexes of Storm·Rainfall Intensity in mm/hr

Event		i_r	$i_r^{(5)}$	$i_r^{(60)}$	$i_r^{(k)}$ moments	LS
1.	Apr 25	2.57	16.8	6.2	3.6	6.8
2.	May 03	2.45	7.7	4.6	3.5	3.4
3.	Jun 28	9.96	20.8	3.9	5.8	6.7
4.	Jul 10	17.92	38.8	9.0	15.4	20.2
5.	Jul 22	6.17	19.8	4.9	2.0	5.4
6.	Aug 10	46.44	74.6	15.5	46.4	46.4
7.	Sep 10	8.34	19.1	11.0	10.5	11.7
8.	Sep 21	10.45	26.9	5.9	14.1	16.4
9.	Sep 24	64.20	94.4	64.2	75.2	76.8
10.	Sep 26	19.60	38.0	4.9	11.8	14.7
11.	Oct 26	1.74	4.7	3.4	2.8	2.9
12.	Dec 18	2.93	5.6	4.5	4.0	4.2

Such observations are confirmed by regression analysis which has been performed to analyze relationships in the form

$$k = a \ x^{-b} \tag{6}$$

relating the time-scale to the examined storm rainfall intensity indexes, taken as the independent variable, x. The results from such analysis are summarized in Table 5.

As it can be observed in Figure 7, the experimental relationship between the time-scale, k, and the peak discharge of storm runoff is quite different from the above one. In fact, the lower bound is detected, suggesting the presence of a limiting value which the time-scale can achieve with increasing peak rates. As a consequence, the power relationship in equation (6) is found to provide a poor fit to the experimental data, as shown in Table 5. A more appropriate relationship should thus account for a threshold level in the form, for instance

$$k = k_o + a q_p^{-b} \tag{7}$$

which has been also investigated through regression analysis. If k_o is assumed equal to 15 min, as suggested by the experimental pattern, the explained variance is found to exceed 87%, with the exponent b ranging from 0.9 to 1.1 depending upon the method of estimation, i.e., moments and least squares respectively.

TABLE 5. Coefficient of Determination and Exponent (in parentheses) from Regression Analysis of the Power Functional Relating k to Storm Characteristics

Independent Variable	i_r	$I_r^{(5)}$	$i_r^{(6)}$	$i_r^{(k)}$	q_p
Moments	0.85	0.66	0.30	0.64	0.47
		(0.52)	(0.57)	(0.46)	(0.47)
LS Estimates	0.86	0.83	0.49	0.86	0.71
		(0.52)	(0.64)	(0.57)	(0.60)

The values taken from the average stream flow velocity, which is related to k through Eqn. (3), are therefore found to be mainly related to the peak flow rate of the runoff hydrograph. Consequently, the runoff hydrograph should reflect the time-variance of the space average streamflow velocity along the channel network during the course of a storm.

The effects of water transport dynamics on the time-scale of the hydrologic response could be synthetized through a monotonic nonincreasing function of the runoff discharge at the catchment outlet, i.e., Eqn. (7). A major implication of this observation is given by the detected presence of the internal nonlinearity involved in catchment response to storm rainfall, resulting as a feedback control effort.

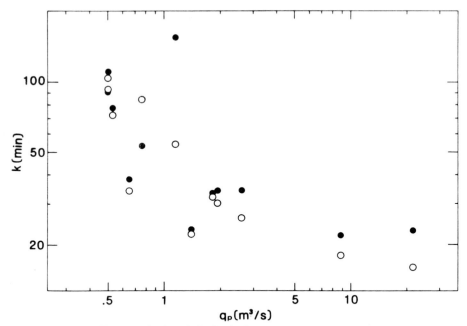

Figure 7   Time-scale, k, of the hydrologic response against peak
discharge of the direct runoff: moments (points) and least squares
(circles) estimates.

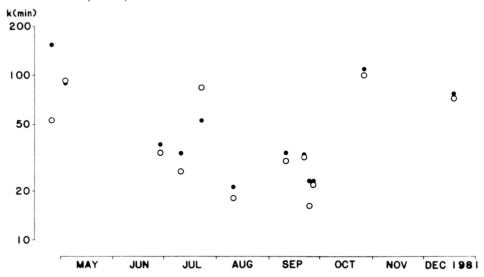

Figure 8   Seasonal variation of the time-scale of the hydrologic response,
k, fitted by the methods of moments (points) and of the least
squares (circles).

This might be reproduced by means of a nonlinear model providing for variable time-scale, which should be related to the system storage and to the output rather than determined from the intensity of the rainfall input.

Notwithstanding the detected complexity of the rainfall-runoff transformation, its nonlinearity is found to decrease with increasing runoff intensity, as it can be observed for events 6, 9 and 10. This might be due to the flow attaining quickly of the bankful capacity throughout the stream network during major floods. The linear assumption is therefore found to increase its reliability with increasing intensity and severity of storm runoff and water transport dynamics might be analyzed in terms of the bankful discharge throughout the stream network in order to estimate the time-scale of the IUH for major floods.

4. DISCUSSION AND CONCLUSIONS

On the basis of the analysis of the storms occurred in Rio della Gallina creek during 1981, nonlinearity is found to affect the rainfall-runoff transformation. Such effects arise from the dependence of the time-scale of the hydrologic response on both the rainfall intensity and runoff flow-rate. A suitable representation of the rainfall-runoff transformation should account for this dependence by introducing models providing for variable time-scale or lag-time of the response function.

Models of the type proposed by Reed et al., (1975), providing for variable lag-time with system storage, seem to provide a better representation of the rainfall-runoff transformation than the one which is achievable through relating the time-scale to rainfall intensity only. The problem of relating system storage to water transport dynamics might be approached by means of hydraulic equations suitable for steep channels, and accounting for hydraulic geometry of bankful flows. In fact, the nonlinearities which has been detected from the external description of the rainfall-runoff transformation may be reflected by the nonlinearities of the internal equations of water transport dynamics in open channels, on one hand, and by the discontinuous boundary which the catchment-stream system provides to the streamflow formation, on the other.

The above presented results come out from a very preliminary analysis and does not allow for any final conclusion. The variability of the time-scale of the hydrologic response of a catchment to storm rainfall appears to be mainly related to storm runoff, as described by its peak flow rate. A **feedback** effect should be involved in such a behavior, of which is in agreement with the satisfactory results achieved by using ARMAX stochastic models in order to infer the rainfall-runoff transformation at the basin scale from input and output observations (Anselmo and Ubertini, 1979).

Notwithstanding the complexity required for an appropriate representation of this transformation, the linear assumption may provide a reliable postulate for major flood analysis also at the elementary basin scale. This result emanates from the decreasing nonlinear effects which have been observed for increasing intensity of the storm event, and it can be inferred to depend on hydraulic geometry of stream networks. This result seems to be also scale-independent, as it has been observed by the authors also for catchments larger than the investigated ones.

Further developments of the research will properly approach the latter item, i.e., the spatial scale problem. The following aspects should be involved:

(a)   comparison of the performances provided by linear time-invariant, linear time-varying and **nonlinear** models;

(b)   parametrization of a nonlinear model providing for variable time-scale as related to the
       space average streamflow velocity throughout the stream network;

(c)   reliability analysis of the linear assumption for major floods with particular reference to
       drainage basin area;

(d)   parameterization of a linear model of major floods using bankful capacity and hydraulic
       geometry of the channels which form the essential component of the stream network of
       a catchment.

ACKNOWLEDGMENTS

The useful discussions with I. Rodríguez-Iturbe were helpful in shaping some of these ideas,
while I. Becchi gave some suggestions for data processing and exchange. Their contribution
has been highly appreciated by the authors.

Thanks are also due to M. Govi, director of CNR IRPI in Turin, for his suggestions and useful discussions of geomorphological aspects involved in this research.

Original drawings were prepared by Mr. E. Viola and processed by Mr. P.G. Trebo, whose
help is gratefully acknowledged, Mrs. T. D'Agostino is thanked for preparation of the final
version of the manuscript.

REFERENCES

Anselmo, V., DiNunzio, F. and Dogone, F., 1982. Hydrological Investigations in a Small
Catchment, *Proc. Symp. Hydrology. Res. Basins,* Sonderh. Landeshydrologie, Bern, 2, pp.
259-268.

Anselmo, V. and Ubertini, L., 1979. Transfer Function -- Noise Model Applied to Flow Forecasting. *Hydrol. Sci. Bull.,* 24: pp. 353-359.

Caroni, E. and Tropeano, D., 1981. Rate of Erosion Processes on Experimental Areas in the
Marchiazza Basin (Northwestern Italy). *Proc. Symp.* "Erosion and Sediment Transport
Measurement," Florence, June 22-26, IAHS Publ. no. 133, pp. 457-466.

Chiu, C. and Bittler, R.P., 1969. Linear Time-Varying Model of Rainfall Runoff Relation.
Water Resources Research, 5(2), pp. 426-437.

Diskin, J.C.I., 1973. The Role of Lag in a Quasi-Linear Analysis of the Surface Runoff System. *Proc. 2nd Int. Symp. Hydrology,* Water Resources Publications, Fort Collins, CO, pp
133-144.

Diskin, J.C. I., 1982. Nonlinear Hydrologic Models. In: Singh V.P. (Editor), *Rainfall-Runoff
Relationship.* Water Resources Publications, Fort Collins, CO, pp. 127-146.

Dooge, J.C.I., 1973. Linear Theory of Hydrologic Systems. *Techn. Bull.,* no. 1468, U.S.
Department of Agriculture, Washington, 327 pp.

Dooge, J.C.I., 1979. Alternative Approaches to Flow Problems. *Proc. XVIII Cong. IAHR,* Cagliari, September 10-14, 2, pp. 28-55.

Gupta, V.K., Waymire, E. and Wang, T.C., 1980. A Representation of an Instantaneous Unit Hydrograph from Geomorphology. *Water Resources Research,* 16(5), pp. 855-862.

Mandeville, A.N. and O'Donnell, T., 1973. Introduction of Time Variance to Linear Conceptual Catchment Models. *Water Resources Research,* 9(2), pp. 298-310.

Nash, J.E., 1960. A Unit Hydrograph Study with Particular Reference to British Catchments. *Proc. Instn. Civ. Engrs.,* 17, pp. 249-282.

Nash, J.E. and Sutcliffe, J.V., 1970. River Flow Forecasting Through Conceptual Models, 1, A Discussion of Principles, *J. Hydrol.,* 10, pp. 282-290.

Pilgrim, D.H., 1983. Some Problems in Transferring Hydrological Relationships Between Small and Large Drainage Basins and Between Regions. *J. Hydrol.,* 65(1/3), pp. 49-72.

Reed, D.W., Johnson, P. and Firth, J.M., 1975. A Nonlinear Rainfall-Runoff Model Providing for Variable Lag-Time. *J. Hydrol.,* 25, pp. 295-305.

Rodríguez-Iturbe, I., and Valdés, J.B., 1979. The Geomorphologic Structure of the Hydrologic Response. *Water Resources Research,* 15(6), pp. 1409-1420.

Rodríguez-Iturbe, I., Gonzales-Sanabria, M. and Bras, R.L., 1982. The Geomorphoclimatic Theory of the Instantaneous Unit Hydrograph. *Water Resources Research,* 18(4), pp. 877-886.

Rosso, R., 1984. Nash Model Relation to Horton Order Ratios. *Water Resources Research,* 20(7), pp. 914-920.

Rosso, R. and Tazioli, G.S., 1979. Suspended Sediment Transport During Flood Flows from a Small Catchment. *Proc. XVIII Congr. IAHR,* Cagliari, September 10-14, 5, pp. 65-72.

Wang, C.T., Gupta, V.K. and Waymire, E., 1981. A Geomorphologic Synthesis of Nonlinearity in Surface Runoff. *Water Resources Research,* 17(3), pp. 545-554.

APPENDIX A - Notation

A	:	drainage area, in sqkm;
a,b	:	parameters of power regression;
ES	:	mainstream slope, in %;
H	:	function arising from the LS solution;
I	:	incomplete gamma function;
i_r	:	average rainfall excess intensity, in mm/hr;
$i_r^{(h)}$	:	maximum rainfall excess intensity with h minutes duration, in mm/hr;
k_o	:	lower bound to k, in min;
k	:	scale parameter of the IUH, in min;
L	:	main stream length, in m;
$\hat{L}_1$	:	mean 1-st order stream length (intercept of the interpolation line with the 1-st order abscissa), in m;
L_1	:	average 1-st order stream length (measured), in m;
L_Ω	:	stream length of Ω-th order channels, in m;
n	:	shape parameter of the IUH;
OLS	:	overland slope, in %:
Q_p	:	total discharge at peak, in cum/s;
q_p	:	direct runoff discharge at peak, in cum/s;
q_b	:	base-flow at the beginning of the runoff event, in cum/s;
q_i	:	runoff discharge at the end of the i-th interval (observed), in cum/s;
$\hat{q}_i$	:	runoff discharge at the end of the i-th interval (computed), in cum/s;
$\bar{q}$	:	mean runoff discharge for a storm event, in cum/s;
R	:	total rainfall depth, in mm;
R_a	:	stream area ratio;
R_b	:	bifurcation ratio;
R_L	:	stream length ratio;
r	:	total rainfall excess depth, in mm;
r_i	:	rainfall excess depth cumulated over the i-th interval, in mm;
S^2	:	objective function for the LS estimate of k;
t_r	:	duration of the rainfall excess event, in min;
t_L	:	lag time, in min;
U	:	finite duration unit hydrograph;
$u(t)$	:	instantaneous unit hydrograph (IUH);
v	:	stream average space-time velocity, in m/s;
x	:	independent variable for power regression;
$\Gamma(n)$	:	gamma function;
ΔT	:	sampling interval, in min;
ϵ	:	residual;
Φ	:	runoff coefficient, equal to r/R;
Ω	:	stream order.

APPENDIX B - Least squares estimation of the time scale of the IUH.

The objective function for the least squares (LS) estimation can be written in the form

$$S^2 = \sum_j \epsilon_j^2 \tag{B-1}$$

where $\epsilon_j = q_j - \hat{q}_j$ and, assuming a constant rainfall intensity during each time interval ΔT

$$\hat{q}_j = \sum_{i=1}^{j} (r_i / \Delta T) \, U_{j-i+1}. \tag{B-2}$$

The objective function is completely specified by letting

$$U_{j-i+1} = I_{j-i+1} - I_{j-i} \tag{B-3}$$

where I_m is the incomplete gamma function

$$I_m = [1/\Gamma(n)] \int_0^{m \Delta T / k} t^{n-1} \exp(-t) \, dt \tag{B-4}$$

For given n, the value of k which minimizes S^2 is such that $d \, S^2 / dk = 0$, that is

$$2 \sum_j \epsilon_j \, d \epsilon_j / dk = 0. \tag{B-5}$$

Moreover, the derivative may be developed as

$$d \epsilon_j / dk = - d\hat{q}_j / dk = \tag{B-6}$$

$$= - \sum_{i=1}^{j} (r_i / \Delta T) dU_{j-i+1} / dk$$

$$= - \sum_{i=1}^{j} (r_i / \Delta T)(dI_{j-i+1} / dk - dI_{j-1} / dk).$$

Finally, having written

$$H_m = dI_m / dk = \tag{B-7}$$

$$= (1/\Gamma(n)) d / dk \int_0^{m \Delta T / k} t^{n-1} \exp(-t) \, dt$$

$$= - (1/\Gamma(n))(m \, \Delta T / k)^n \exp(-m \, \Delta T / k),$$

hen

$$d \epsilon_j / dk = - \sum_{i=1}^{j} (r_i / \Delta T)(H_{j-i+1} - H_{j-i}) \tag{B-8}$$

The optimization may then be achieved by some numerical method for solving nonlinear equation (B-5). The "regular falsi" has been adopted for this purpose, since it has been found to provide smoother and more reliable convergence to the solution, as compared with the performances achievable by means of the Newton-Raphson algorithm.

3

A RUNOFF SIMULATION MODEL BASED ON HILLSLOPE TOPOGRAPHY

Mike Kirkby

ABSTRACT

The detailed model presented is a development of an exponential store model (TOPMODEL) previously described (Beven and Kirkby, 1979). The continuity equation for saturated sub-surface flow is expressed in terms of unit runoff which emphasizes the relatively low spatial variation in rates. For this reason kinematic wave solutions have not been used to solve the partial differential equation. A single unsaturated and a saturated store are considered at each point down the hillslope length, the former delaying infiltration and the latter providing downslope sub-surface flow and establishing saturated contributing areas.

The total sub-surface flow at saturation, slope gradient and hillslope plan form are used to generate flow differences down the length of the hillslope profile. Simulations show the generation of saturated overland flow at downslope sites where flow converges, superimposed on a hydrograph which is largely controlled by the convex (in profile) divide area. This suggests that for most natural slopes, the runoff delivered to channel banks may be estimated efficiently from two separate linked component models. The first forecasts spatially uniform flow at rates determined by topography and soils in the hilltop divide areas. This model is able to forecast the changing saturated area on which the second component model forecasts the saturated overland flow. This or another appropriate hillslope flow model may be combined with a flow routing algorithm for the channel network to give catchment hydrological response. In simple cases of spatially constant network routing velocity, the resulting combined model is identical to the *Geomorphological Unit Hydrograph* of Rodriguéz-Iturbe, providing an explicit physical meaning for his theoretical unit response functions as the hillslope base unit hydrograph.

1. INTRODUCTION

In the absence of rainfall and runoff records for a particular catchment, runoff estimation relies on comparison with data from other catchments. Where, as is usual, the comparison is not exact, corrections are needed on the basis of meteorological and drainage basin

39

V. K. Gupta et al. (eds.), Scale Problems in Hydrology, 39–56.
© *1986 by D. Reidel Publishing Company.*

characteristics. These corrections may rely wholly on empirical correlation between instru-
mented catchments, or may depend to some extent on physical models for runoff generation.
The advantages of a model approach are that it may suggest novel forms of catchment
parameters which are not evident from correlation; and that parameters reached through a
model are likely to be more reliable under modest extrapolation beyond the data. These
advantages will, however, only be realized in a useful way if the models developed are rather
simple, so that the relevant parameters may be both comprehensible and not too many. An
additional reason for using a simple model is that hydrographs show only a limited range of
forms, and the number of independent model parameters should not greatly exceed the
dimensionality of hydrograph form if its parameters are to be sensibly determined.

The model explored here is a development of the exponential store model described by Beven
and Kirkby (1979) and tested on UK catchments of up to $10km^2$ (Beven, et al., 1984). In the
simplest form of this model, both sub-surface and overland flow is generated from a single
saturated store which is considered to extend downwards indefinitely into the ground and
reach up to a level which ranges over time and with position. Zero storage is defined as the
condition of saturation to the ground surface, so that negative storage values are associated
with saturated sub-surface flow and positive values with overland flow (together with a maxi-
mal rate of sub-surface flow). If net rainfall is added directly to such a store, then hydrograph peaks occur immediately at the end of simple storms. An unsaturated store is commonly added to the minimal model, serving to delay percolation into the saturated store. It
has been assumed that there is no lateral unsaturated flow. The unsaturated store may also
be used to forecast infiltration capacity and Hortonian overland flow, and to correct potential
to actual rates of evapotranspiration. The form of unsaturated store used here most closely
follows that described in Beven and Wood (1983).

In this paper the model equations are set out in terms of runoff rates, for the general case of a
hillside flow-line strip. Topography and soils are specified in terms of gradient, flow-line
width (allowing flow con- or di-vergence) and saturated lateral permeability. Other soil/slope
parameters have been held constant along the strip. Flow is routed downslope through distributed unsaturated and saturated stores to generate slope-base and hillside hydrographs in
response to storm rainfalls. The behavior of this kind of model is explored below, considering
apropriate simple forms for the unsaturated store(s), and examining the behavior of the model
in response to changes in its main parameters. The first purpose of this exploration is to
determine what hillside parameters have the greatest influence on hydrograph form. The
second is to consider whether two dominant parameters provide a sufficient basis for a still
simpler model which may then be incorporated into a whole-basin model to combines the
influence of many flow-line strips interacting with a channel network.

2. MODEL EQUATIONS

Flow-line strips are defined as following lines of greatest slope, orthogonal to elevation contours. Distance along the strip, x, is measured from the divide in a horizontal direction following the local flow-line direction. The width of the strip is defined by its width, w, at each
point along it. Drainage area per unit strip (i.e., contour) width, is described by the geometrical relationship:

$$a = \int_0^x x \cdot dx \qquad (1)$$

The geometry of the strip may also be described by two other geometrical identities which are
derived here for use below. From (1):

$$\int_0^x w \cdot dx = aw \qquad \text{or} \quad w = d(aw)/dx \qquad (2)$$

Also from (1):

$$1/a = w/\int_0^x w \cdot dw \quad \text{or} \quad \int (dx/a) = \ln(\int_0^x w \cdot dx) + \text{constant}$$

$$\exp[\int (dx/a)] = \int_0^x w \cdot dx = aw. \tag{3}$$

Along a flow strip, the continuity or storage equation for conservation of water mass may be written:

$$\partial(qw)/\partial x + \partial(Sw)/\partial t = iw \tag{4}$$

where

  q is the local saturated discharge per unit width,
  S is the local saturated water storage (defined to be zero at the ground surface and negative for deficits below saturation),
  t is elapsed time
  and i is the local rate of percolation to the saturated zone.

Replacing discharge by runoff (i.e., discharge per unit area), j, using the geometrical relationship $q = ja$, the continuity equation may be rewritten as:

$$\partial(ajw)/\partial x + \partial(Sw)/\partial t = iw. \tag{5}$$

The central assumption of the TOPMODEL formulation is that the saturated sub-surface store has an exponential flow law, which provides a satisfactory fit to catchment storm response in a range of circumstances (e.g., Beven et al., 1984) and also provides spatially uniform runoff from spatially uniform inputs to the saturated layer (Beven and Kirkby, 1979), as may be seen below. The assumed exponential flow law for saturated flow at any site may be written in the form:

$$q = aj = q_o g \exp(S/m) \quad \text{or} \quad S = m \ln\{j \cdot [a/(q_0 g)]\} \tag{6}$$

where q_0 is the saturated soil discharge on unit hydraulic gradient, and the scaling soil parameter, m is assumed to be spatially uniform along the strip. Expanding the first term in equation (5) and substituting for S, it may be seen that j is the only time-dependent term in the partial differentiation with respect to time, so that, with the exponential store assumption, the continuity equation may be re-written as:

$$j\,\partial(aw)/\partial x + aw\,\partial j/\partial x + (m/j)\partial j/\partial t = iw$$

Substituting the geometrical relationship expressed in equation (2), rearranging terms and dividing through by w:

$$a\,\partial j/\partial x + (m/j)\partial j/\partial t = i-j \tag{7}$$

This expression provides a basis for routing saturated flow down the length of the hillslope strip, and is used below for the computer simulations.

Two special cases may readily be solved analytically; for a steady state over time and for spatially uniform runoff under various assumptions. For the steady state $\partial j/\partial t = 0$ by definition, so that the flow equation reduces to the ordinary differential equation:

$$adj/dx + j = i \tag{8}$$

Using the integrating factor $\exp(\int dx/a)$, and the identity (3);

$$j = [1/aw] \int (iw) \cdot dx \qquad (9)$$

For a spatially uniform input, i then runoff $j = i$. For other steady inputs the result is non-trivial, though it may be derived more readily direct from continuity considerations.

The more interesting special case is of spatial uniformity, defined by $\partial j / \partial x = 0$. In this case the ordinary differential equation is:

$$dj / dt = (j/m)(i-j) \qquad (10)$$

For the case of zero input, this gives the recession curve:

$$j = j_0 / (1 + j_0 t / m) \qquad (11)$$

where j_0 is the runoff at time zero, which is an adequate fit to many observed catchment regression curves. For constant non-zero input at rate i:

$$j = i / [1 + (i/j_0 - 1)\exp(-it/m)] \qquad (12)$$

For a short burst of percolation, totalling R, the final runoff and the average during the burst are respectively:

$$j = j_0 \exp (P/m) \text{ and } \bar{j} = [mj_0/R][\exp(R/m)-1] \qquad (13)$$

Equation (7) may be solved as a kinematic wave equation, with a wave velocity of aj/m. For the more limited assumption of spatially uniform inputs to an initially nonuniform runoff distribution, the solution shows a wave of nonuniformity passing down the flow-line strip, and being replaced by spatially uniform runoff. Once this state has been achieved, no subsequent addition of spatially uniform percolation can disturb it. With spatially nonuniform inputs of percolation to the saturated layer during a storm, the solutions show convergence towards spatially uniform runoff as soon as percolation has become spatially uniform. This condition is generally reached during flood recessions after rainfall stops. The time to reach near-zero percolation is controlled by the unsaturated store. Subsequently spatial differences are carried away at a rate given by the kinematic wave velocity above, which normally suggests survival times of only a few hours. For this reason, a kinematic wave solution has not been used here, reflecting the poor development of visible wave phenomena.

3. THE SATURATED STORE

The most important soil parameter for the hillslope flow strip is the soil constant, m. By making some assumptions about the saturated soil behavior, values for m may be linked to more readily measured soil parameters, and in particular to the profile of lateral saturated permeability. Observed velocities of downslope subsurface flow commonly require effective lateral permeabilities of the order of 100 meters per hour, which can only plausibly be supported if most of the flow occurs in relatively large connected systems of macropores. It is here assumed that **all** of the lateral flow occurs in macropores, which are assumed for simplicity to form a network of linear cracks, at equal frequency and of decreasing diameter with depth. A microporosity is also assumed to be present, supporting most of the saturated storage but none of its flow. Effective microporosity is also assumed to decline with depth, and to be related to the decline in macroporosity: here simple proportionality is assumed as a basis for proceeding. With these assumptions the relationship between saturated storage level S and depth in the soil z may be established, together with a form for the associated permeability profile with depth.

For laminar flow in a linear crack of diameter d, the mean flow velocity is given by:

$$v = [\rho g / (6\eta)] d^2 = d^2/\alpha \qquad (14)$$

where

ρ is water density,

g is the gravitational acceleration,

η is the kinematic viscosity of the water and $\alpha = 6\eta/(\rho g)$.

The mean flow velocity for all the water is obtained by correcting by the ratio of macroporosity ϵ to total porosity ϵ'. This mean velocity is also given by the marginal velocity dq/dS obtained from the form of the saturated store in equation (6) above for unit gradient ($g = 1$), so that;

$$[dq/dS]_{g=1} = (q_0/m)\exp(S/m) = (\epsilon'/\epsilon)d^2/\alpha \tag{15}$$

If macropore cracks occur at constant frequency of n per unit length, then macroporosity is given by:

$$\epsilon = nd \tag{16}$$

By definition, the marginal change of storage with depth is equal to the total available porosity ϵ', so that, applying (16):

$$\epsilon' = -dS/dz = \lambda nd \tag{17}$$

where

λ is the ratio ϵ'/ϵ of total to macro-porosity.

Substituting for d from (15):

$$\epsilon' = \lambda n(\lambda\alpha q_0/m)^{1/2}\exp[S/(2m)] = \epsilon_0' \exp[S/(2m)] \tag{18}$$

where $\epsilon_0' = \lambda\rho(\lambda\alpha q_0/m)^{1/2}$ is the total porosity at the soil surface.

Integrating over depth, the relationship between depth and storage is:

$$z = (2m/\epsilon_0')\{\exp[-S/(2m)]-1\} \tag{19}$$

and the saturated lateral permeability K_m is:

$$K_m = (\epsilon' q_0/m)\exp(S/m) = (\epsilon_0' q_0/m)/[(\epsilon_0' z/2m)+1]^3 \tag{20}$$

Equation (20) shows a decline in permeability with depth at an initial proportional rate of $3\epsilon_0' z/(2m)$. The form of the decline may be compared with that fitted to a number of soil profile data by Beven (1983), using the form:

$$K_m = K_0\exp(-fz) \tag{21}$$

Comparing rates of decline near the surface, it may be seen that $f = 3\epsilon_0'/(2m)$. Estimated quartile values of $1/f$ (Beven, 1983, Appendix) for 27 soils lie at about 200 and 400 mm. If available porosity for saturated soil at the surface is 5%, then this range of values corresponds to m-values of 15 to 30 mm. These values may be compared to those estimated for UK experimental catchments of 5 to 40 mm, although the lowest of these m-values is for soil with a much higher clay content than any of these analyzed for permeability. It is also worth noting that measured values for permeability generally underestimate the contribution of macropores, so that only order of magnitude comparisons should be attempted from such data.

This analysis may be extended to estimate the maximum rate of percolation into the saturated zone from the soil above. The result of this analysis is not reported here in detail but relies on an extension of the approach already outlined. Rates forecast in this way are at least an order of magnitude higher than observed rates of percolation. It is suggested that percolation downward is effectively controlled entirely through the possible rate of outflow from the unsaturated zone, which is considered next.

4. THE UNSATURATED STORE

Flow in the unsaturated zone may be modelled in detail using the unsaturated flow equations in conjunction with characteristic curves for soil moisture *versus* hydraulic potential (with or without hysteresis). Treating the unsaturated zone as a single store is inevitably far from perfect, but may be adequate given the low dimensionality of hydrograph forms. The imperfections are greatest where a given total moisture content may be associated with very different vertical moisture distributions. An extreme example is the difference between conditions in a capillary fringe in equilibrium with a static water table, and conditions of rapid infiltration into a dry soil with moisture concentrated near the surface. In the former case there is zero downward flow, gravity balancing hydraulic potential gradients; whereas in the latter case the two gradients combine to maximize downward percolation. In a conceptual single store, percolation must be estimated as an average lying between such extremes, normally for the case of zero hydraulic gradient. Flow is then approximately equal to the vertical permeability, which depends primarily on moisture content. On this view the downward percolation should depend on the ratio $h/(-S)$ is equal to the proportion of moisture saturation in the zone above that of saturated flow. For simplicity and because it appears to give the most reasonable results, a linear dependence has been used in modelling, even though the relationship between permeability and moisture content is generally found to be more than linear. It has also been assumed throughout that the unsaturated zone is one of downward percolation only, with no lateral component.

An alternative view of the unsaturated store is as a means of delaying the hydrograph peak. From this point of view, percolation at a rate directly proportional to $h/(-S)$ is equivalent to treating the unsaturated store as linear, with a delay time proportional to the saturated deficit, $-S$. Both points of view appear to give a reasonable direction of dependence on the variables involved, and follow the model of Beven and Wood (1983). Thus greater deficits increase the delay to hydrograph peak following rainfall, and greater unsaturated storage levels give greater percolation for a given saturated deficit. The relationship is however likely to break down for low positive deficits, and certainly does so for negative deficits. It is proposed here that at low deficits, the maximum rate of percolation is controlled either by the absolute unsaturated storage level, h, or by the rainfall intensity if that is higher.

The argument for using total storage as a determinant of percolation may be related to the condition of maximum percolation described above, namely of rapid infiltration above an advancing wetting front. In that case downward percolation from the front responds to the depth of the front directly, moisture content above it lying close to saturation. Again a linear proportionality has been found to be suitable. On the alternative view of the unsaturated store as linear, an upper limit proportional to h is equivalent to a minimum time for drainage. Plainly if the drainage time is able to decrease indefinitely then percolation outflow may in some circumstances greatly exceed storm rainfall intensity, a situation which is rarely if ever encountered in practice. When rain is falling on a saturated or almost saturated soil, then its intensity is also thought to be an appropriate upper limit for percolation if it is higher than the percolation estimate just described. In this case the unsaturated zone is acting simply as a shunt to transmit rainfall pulses directly.

To maintain continuity between percolation at a rate proportional to $h/(-S)$ and percolation at a rate proportional to h, the critical transition must occur at a fixed deficit, which may sensibly be related to the soil scale depth parameter m. Where rainfall intensity sets a greater upper bound, then the transition occurs at a lower deficit. This version of the unsaturated store has been adopted.

The unsaturated store may also be used in a model to limit transpiration rates and to determine infiltration capacity. Transpiration may for example be allowed to exhaust all or part of the unsaturated store, and then draw the saturated store down to some maximum deficit. The adoption of an abrupt cut-off of this sort is thought adequate at a single site, and will produce a progressive decline in transpiration rates when applied to a drying catchment area, following the areal distribution of deficits. Infiltration capacities may similarly rely on the unsaturated storage level, using the Green and Ampt (1911) equation which is based on storage, with a 'steady' leakage rate based on the downward percolation discussed above. Excess rainfall may then be routed overland at the same marginal rate as for saturated overland flow, q_0/m. These facets of the model have been adopted in principle, but not critically tested.

The conceptual model adopted for the unsaturated store is formally described by the following equations:

$$i = \beta h / (-S) \tag{22}$$

valid for large deficits,
where i is the rate of percolation into the saturated store,
and β is a percolation rate parameter (e.g., in $mm.hr^{-1}$)
For small deficits,

$$i = \beta h / (\gamma m) \tag{23}$$

where γ is a constant parameter.

This expression is valid for deficits less than γm, where γ is thought to be of the order of unity. At high rainfall intensities, equation (23) is replaced by:

$$i = r \tag{24}$$

where r is the rainfall intensity, valid when $r > \beta h / (\gamma m)$.
The infiltration capacity f may be estimated as:

$$f = i + A / h \tag{25}$$

for constant A, following Green and Ampt (1911).
Analysis of this equation for infiltration into a soil with large saturated deficit and $h = 0$ initially gives, with no saturated outflow, an initial rapid decrease in f *prop* $t^{-1/2}$, followed by an eventual very gradual rise *prop* $t^{1/8}$. If the deficit is held constant during infiltration the initial fall is followed by a sharper eventual rise *prop* $t^{1/2}$. These relationships depart somewhat from standard infiltration theory which requires an eventual constant infiltration rate, but the departures appear to be acceptably small in most cases.

5. GENERAL MODEL BEHAVIOR

Figure 1 is a schematic flow diagram for the water balance of each segment along the flow strip modelled. Equation (7) has been solved computationally, using a simple explicit scheme on a micro-computer. Stability has been maintained by calculating' rates of change of saturated runoff, and reducing the time step so that this change nowhere exceeds a threshold proportion (usually 0.01). Initial conditions are set by calculating equilibrium with an assumed steady rate of rainfall and transpiration. In the simulations described below, the potential transpiration rate has been set to zero, and the rate of infiltration to exclude Horton overland flow (by setting the constant A in (25) high). Unsaturated and saturated storages are calculated for up to 20 downslope segments, each one specified by its width and gradient $\times$ permeability. Initial conditions have been obtained by assuming equilibrium with a steady rate of antecedent net precipitation, and an equal rate of percolation and runoff at every point.

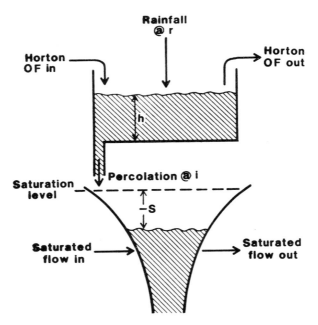

FIGURE 1. Schematic flow diagram for hillslope runoff model.

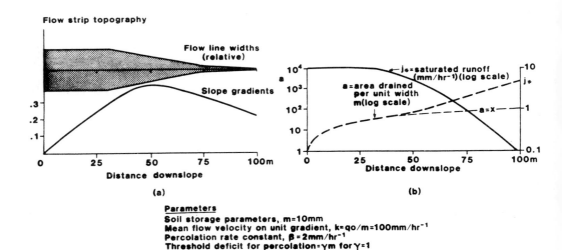

FIGURE 2. Standard flow strip topography adopted in model runs.
(a) Relative flow-line widths and gradients assumed
(b) Resulting areas drained per unit strip width (m.) and saturated runoff rates $(mm.hr^{-1})$.

Simulations have been made with a standard set of parameter values and topographic form, with a simple storm at the start of the simulation period. The topographic form is shown in Figure 2(a), which illustrates a convexo-concave profile on a flow strip which shows strong convergence of flow near the base of the hillslope. The form is representative of a valley-head hollow in a temperate environment. The area drained per unit contour length, a is shown in Figure 2(b). It may be seen to be initially equal to distance from the divide and then to increase rapidly, reaching a value of 2,355 m at the base of the slope, due to the effect of flow convergence. Figure 2(b) also shows the runoff at saturation, $j_* = q_0g/a$, for the adopted value of q_0. The effect of flow convergence and profile concavity reinforce one another at the base of the slope, so that saturation occurs at a runoff of 0.089 $mm.hr^{-1}$ at the slope base and at 10 $mm.hr^{-1}$ near the divide. The relevant 'standard' parameter values, antecedent runoff and storm size are listed in Table 1.

TABLE 1: Standard Parameter Values Adopted in Model Runs

Soil storage parameter, $m = 10$ mm [Equation 6]
Mean saturated flow velocity on unit gradient, $q_0/m = 100m.hr^{-1}$ [Equation 6]
Percolation rate constant, $\beta = 2mm.hr^{-1}$ [Equation 22]
Threshold deficit for percolation $= \gamma m$ for γ 1 [Equation 23]
Rain storm at $20mm.hr^{-1}$ for 3 hours
Initial runoff at $0.5mm.hr^{-1}$

For these standard conditions the main hydrograph peak occurs about 8.5 hours after the start of the 60 mm storm. The form of this standard response is illustrated in Figure 3(b). The upper part of the slope (0-30 m from divide) generates the solid curve labelled '0.5', while the slope base generates the broken-line curve labelled '0.5'. It may be seen that the saturated hollow area generates a quick response, which is superimposed on a larger sub-surface response from the slope as a whole. The results are thought to represent flood response from a hollow with moderately permeable soils.

It has not been practicable to explore the complete parameter space, but simulations have followed changes in one parameter at a time about the standard values described above. Figure 3(a) shows the response of the upper part of the hillslope to changing storm volume (60 mm as standard). Storms have all been evenly distributed over a three-hour period, although changes in the distribution have only a limited influence in the overall hydrograph form. As may be expected, the responses show moderate nonlinearity, with lag times decreasing, and peak flows increasing more than in proportion to storm volume. At storm volumes above 200 mm the peak coincides with the end of the storm. For storms of less than about 20 mm the trend in lag times is reversed, until for storms less intense than the antecedent runoff the lag is zero.

Figure 3(b) shows the response of the divide area (solid curves) and the slope base (broken lines) to changes in antecedent runoff. (0.5 $mm.hr^{-1}$ as standard). Again lag times decréase and peak flows increase for wetter conditions. At the slope base the quick response peak also becomes increasingly important, although it inevitably peaks at the end of the storm. In each case it may be noted that the slow response peak from the slope base is in time with the upslope peak, and this occurs for all intermediate points.

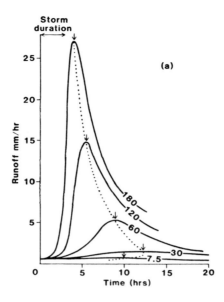

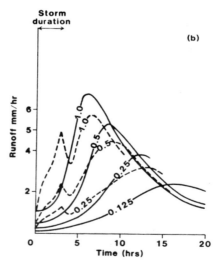

FIGURE 3. The influence of storm rainfall and antecedent conditions on simulated hillslope floods.

> (a) Effect of storm rainfall (60 mm. standard) on convex upper slope: all storms spread over a 3-hour period. Arrows and dotted line indicate peaks.
> (b) Effect of antecedent equilibrium runoff ($0.5mm.hr^{-1}$ standard) on response to 60 mm. storm. Solid line for upper convexity: broken line for slope base.

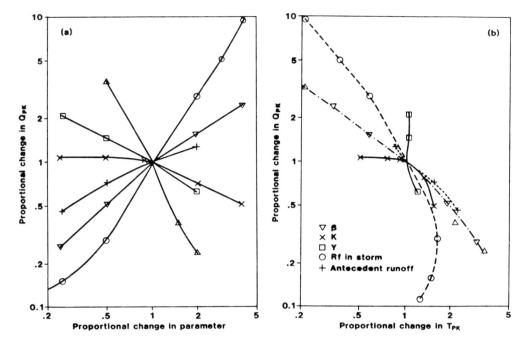

FIGURE 4. The influence of parameter values on modelled hydrograph form as one parameter is changed at a time.

(a) The proportional response of the slow hydrograph peak to proportional changes in parameter values.
(b) The proportional changes in peak flow and lag to peak from storm centroid as parameter values are changed.

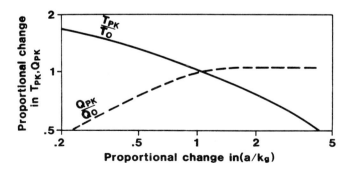

FIGURE 5. The influence of topography on peak flow and lag to peak. Proportional changes are shown in response to proportional changes in the topographic scale time, $t_G = a/(kg)$.

Changes in parameter values generally influence both the magnitude of the hydrograph peak and its timing, as well as having some influence on other aspects of hydrograph form. Figure 4(a) shows the proportional change in peak flow for proportional changes in each of the major model parameters. Storm volume and antecedent runoff are also included for comparison. It may be seen that the saturated permeability $k = q_0/m$ has least influence on peak flows, and that k, the soil parameter m and the percolation threshold parameter γ all reduce the peak as they increase; whereas peak flow increases with antecedent runoff j_0, storm volume R and the percolation rate parameter β. The figure is plotted for the upslope response where this is delayed beyond the end of the storm. In some cases the change to a quick response at extreme parameter values leads to a discontinuity in the relationships shown, as the peak flow adopts a value related to peak rainfall intensity.

As may be expected from considerations of total runoff volume, increases in peak flow for a given storm volume generally lead to decreased lag time. Figure 4(b) shows the relationships between proportional changes in simulated peak flow and time to peak (from the storm centroid) as parameters are changed, one at a time. It is evident that the form of the hydrograph can be changed most readily through the parameters k, which change lag more than peak flow, and γ which changes peak flow more than timing. The parameters β and m both have a similar effect on lag as they change the peak flow. However changes in m tend to reduce the quick and slow peaks roughly in proportion whereas changes in β reduce the slow peak faster than the quick peak, which therefore gains in relative importance. Thus opposing changes in m and β change the ratio of quick to slow peak flow for a given time lag.

The effect of topography is somewhat more complex than the response to single parameters because it is represented by the two values of gradient $\times$ permeability and width for each slope segment, providing 40 parameters in all. They have the advantage in a forecasting sense that they may, ignoring permeability differences, be rather unambiguously measured for a particular site, but that does little to explain their influence on hydrograph form. It has already been shown above that for the topography used in the simulations, the hydrographs have a slow peak which is almost perfectly synchronous all down the slope. It is worth noting here that for slopes where $a \approx s$ along the profile, the storage levels, both saturated (S) and unsaturated (h), are spatially constant at all points, although changing through time. This condition is met for a number of topographic forms, of which the most immediately relevant is that of a simple convex profile (gradient $\approx$ distance from divide) combined with a ridge, spur or col plan-form on which the strip width, $w \approx x^{\lambda}$ at distance x from the divide for constant λ. The unit area a is then given by:

$$a = x/(1+\lambda) \quad \text{for } \lambda > -1 \qquad (26)$$

This class of slopes includes the convex ridge at the top of the standard model slope. It should also be noted that the parameter $k = q_0/m$ scales saturated flow velocities in exactly the same way as gradient, so that changes in k produce identical effects to overall proportional changes in gradient.

Figure 5 shows the influence of the ratio $a/(kg)$ on peak flow and lag, for the cases where there is a distinct slow-response peak, and for slopes or parts of slopes where the ratio is constant from the divide. This ratio, which depends on topography and to some extent on soils, has the dimensions of time and is referred to as the "topographic scale time", denoted by t_G. For other topographic forms it is very clear that local values of t_G no longer give the same functional dependence of time lag on topographic factors. For slopes with an appreciable convex element near the divide (within which t_G is constant), the slow peak is seen to be very strongly influenced by the near-divide timing even where the form of the slope base departs very markedly from constancy of t_G (linear convexity), as for example with the standard slope form used in simulations. Retaining the "standard" convexo-concave profile and changing the plan form of the flowstrip, dominance by the divide convexity (and its t_G value) remains strong for a strip of uniform width (a ridge side-slope), but the slope-base hydrograph shows the influence of more local (i.e., slope base) topography where the flow strip forms a spur (with increasing strip width downslope). Examination of these and other topographic forms suggests that an integrated value of t_G for comparison with the simplex convex slope case can best be calculated as a weighted mean of $a/(kg)$ values from the divide to the point of interest. An appropriate empirical weighting factor consists of the strip width squared, although no rationale has been developed to support this value. Using this weighting, it may readily be seen that local slope base topography is much more influential for spurs than for hollows or straight slopes in determining slope base hydrographs.

6. "QUICK" AND "SLOW" HILLSLOPE RESPONSE AT THE BASIN SCALE

The analysis of topography above for a distributed flow strip allows approximations to be made about the performance of the whole strip as a lumped unit. A two-component model may be obtained by separately modelling the "slow" and "quick" responses. The "slow" response is used here to indicate the hydrograph peak which occurs *after* rainfall has ceased for a simple storm, which is largely produced by subsurface flow. The "quick" response indicates the (usually) secondary peak reached *at the end of* a simple storm, which generally has a substantial saturated overland flow component. The topographic scale time may be used to re-write the storage-flow relationship of equation (6) in the form:

$$j = (m/t_G)\exp(S/m). \qquad (27)$$

Together with the other parameters m, β and γ or their average values for the flow strip, it may be used to generate the subsurface flow peak (or the main flow peak if there is no slow peak) for the slope base. The procedure is identical to that for the distributed model except that only a single point is required. The slope is treated as if it were simply convex with the average value of t_G describing the convexity.

The amount of runoff in the quick response peak may be calculated from the area contributing runoff at the current rainfall intensity. Figure 6 illustrates the necessary analysis of the topographic data. The structure of the unsaturated store requires that rainfall is transmitted directly to the saturated store if the saturated deficit is less than γm, so that the contributing area at any time is that for which the current runoff $j > j_* \exp(-\gamma)$, where $j_* = m/t_G$ is the local runoff at saturation. Figure 6 illustrates the form of the dependence for a number of the topographic forms used in model simulations. The current contributing area may be read directly in terms of the current runoff level, calculated using the slow runoff component described above. At each time interval, the volume of rainfall supplying quick runoff is equal to the current rainfall intensity multiplied by the contributing area. This estimate, summed over a storm, gives an estimate of total quick runoff contribution which compares very favorably with that obtained for the distributed flow strip above, although not all the rainfall contributed in this way flows out of the slope immediately. Where the quick response is large, it may be necessary to iterate this process, using a reduced rainfall amount to generate the slow response, etc.

This description of a two component model is very similar to that described by Beven and Kirkby (1979) and Beven and Wood (1983), but advances it appreciably in two directions. The first of these is in showing that the simpler model does in fact simulate the behavior of the distributed flow strip model with reassuring fidelity. The second, and more important value of this exploration has been in making the form of topographic dependence much more explicit in the simple model, particularly for the slow response peaks.

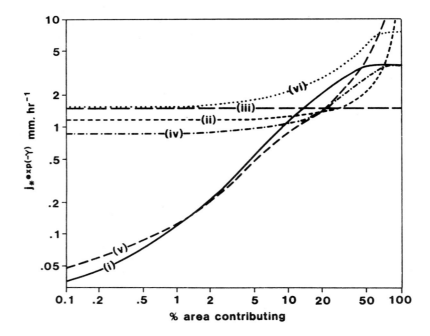

FIGURE 6. The relationship between local runoff rates and contributing area for example
flow strips of differing topography:

        (i) Standard slope

       (ii) Straight (constant gradient) profile on ridge (constant width)

     (iii) Convex (uniformly increasing gradient) profile on ridge

     (iv) Standard profile on ridge

      (v) Straight profile on standard hollow

    (vi) Standard profile on spur (linearly increasing width)

Topography is in fact rather less variable than the range of forms explored above. In most landscapes true slope profiles are dominated by convexities. This is perhaps self-evident for humid-temperate landscapes, but is also true for the typically much shorter slopes of semi-arid lands. Impressions of long concave slopes are commonly illusory, a visual impression being obtained by looking across interfluves which mirror the concavity of the densely spaced intervening channels, as has been noted by Carson (Carson and Kirkby, 1972). True slopes are typically short and more than 50% convex in profile. The predominance of such forms is a necessary conclusion for landscapes of net fluvial erosion, since long concavities are inherently unstable in the sense of developing into permanent eroded channels, following the argument of Smith and Bretherton (1972). Thus for most natural slopes at least two thirds of their length is convex, with a genuinely constant topographic scale time, and this convex divide area consequently dominates the (slow) hydrograph timing and magnitude.

At the basin scale, the combination of flow strip hydrographs may be achieved over an area of constant storm intensity through the distribution of data on divide convexity, which also tends to show some consistent pattern within a catchment. These data are readily obtained from maps or air photographs, etc., and provide one important source of variability on the basin scale. Variability can also be considered in flow strip parameters, which have for sim-plicity (and in the case of m from necessity) been held constant within a strip but which may vary between strips. These data are much less readily available than topographic data and it is worth exploring the degree of explanation which can be achieved through topography alone. For example headwater hollows within the first order humid catchments commonly show lower gradients and more gentle divide convexities than side-slopes within the same catch-ment. This difference in topography is enough to explain both a more rapid slow response and a greater proportion of quick response from the hollow area than from the side slopes. Field evidence can indicate whether these topographic effects are great enough to explain the observed differences of these types, or whether a part of the differences must be attributed to changes in other parameters.

7. INTEGRATION OF HILLSLOPE AND NETWORK RESPONSE

The effect of channel delays on hillslope hydrographs generated according to this or alterna-tive models is well documented. Surkan (1969) shows that the basin output hydrograph takes the form:

$$Q(t) = \sum q(t-\tau)\Delta l(\tau) \tag{28}$$

where

$Q(t)$ is the catchment discharge at time t,
$q(t)$ is the hillslope outflow hydrograph per unit bank length at time t,
and $\Delta l(\tau)$ is the increment in channel kinematic wave travel distance in time $\Delta\tau$.

The summation is made for all positive τ and across all links of the channel network. Beven (1979) has shown that in some circumstances at least, the catchment routing velocity, c may be assumed constant over space and time. It may also be assumed, at least initially, that hillslope outflow is spatially uniform, so that $q(t)$ is independent of location in the catch-ment. With these assumptions, the catchment outflow takes the simple form:

$$Q(t) = c \sum q(t-\tau) \, W[c(\tau)]\Delta\tau \tag{29}$$

where $W(x)$ is the network width at distance x above the outlet, in other words the number of channel links at x, considering all possible upstream paths. The summation is over all positive τ.

It has been shown (Kirkby, 1976) that the catchment hydrograph peak, with these simplifying assumptions, is given with a good degree of approximation by:

$$J_{PK} / j_{PK} = \Phi(z) \qquad (30)$$

where J_{PK}, j_{PK} are respectively the catchment and hillslope peak runoff rates, and the dimensionless parameter $z = uW_*/L$ where $u = c \times$ (time to slope hydrograph peak), $W_* =$ maximum network "width" as defined above, and $L =$ total length of all channel in network. The function $\Phi(z)$ behaves like z for small z, and $\to 1$ for large z. Thus large catchments, in which network travel time is long compared to slope travel time, have a hydrograph which is dominated by network shape and size. Small networks with brief travel times, on the other hand, generate hydrographs which depend strongly on the hillslope input. On the basis of typical slope hydrograph forms, it is thought that the change-over in hydrograph form normally occurs in catchments of between 50 and 250 km^2. It is also demonstrated (*ibid.*, p. 208-211) that for networks of Shreve magnitudes up to 1000, there remains a considerable influence of catchment shape on network and therefore hydrograph form, with a persisting range of about 4 $\times$ for a given drainage area in both hydrograph peak and in delay to peak. It is also implicit in this analysis that basin hydrographs are increasing linearly with respect to storm amount for catchments of greater than threshold size, since the routing velocity is relatively constant, whereas the hillslope hydrograph has been seen above (and demonstrated in the field) to be highly nonlinear. The model presented here is thus consistent with the empirical success of the unit hydrograph approach for large catchments, and its marked failure for small drainage basins.

It should be remembered that considerable simplifications have been made here, although the broad conclusions are thought to be valid with more realistic assumptions. One significant factor which has not been considered here is the spatial distribution of rainfall, which is recognized as having a considerable impact as the ratio of catchment area to storm area is altered.

The form of equations (28) and (29) is almost identical except for notation to the formulation of the geomorphological unit hydrograph (GUH: Rodriguez-Iturbe and Valdes, 1979; Gupta et al., 1986), especially in its most recent form. It is therefore argued that hillslope hydrological response provides the physical mechanism underlying the unit response function of the GUH. There thus appears to be a strong case for investigating all aspects of the catchment routing/GUH model, to establish the practical importance of hillslope flow generation at a basin scale in comparison to the translatory and dispersive influences of network flow.

8. CONCLUSION

The model presented here is seen to offer some advance on existing versions of TOPMODEL, though developed from them. It allows either simple distributed modelling based on hillslope flow strips or explicit forms for integrating topography into a lumped flow strip model, particularly when slopes have substantial upper convexities, as is normal. Aggregation to the basin scale is then readily achieved through the distribution of topographic and soil parameters.

There is also seen to be a real possibility of estimating runoff generated within arbitrary grid squares, for linking with atmospheric circulation or other large-area models. A suitable procedure for a single grid square is thought to consist of first selecting a number of points at random (subject to a suitably stratified design) from the square. It is assumed here that non-topographic parameters (e.g., β, γ and m) are constant within the grid square, and a more complex procedure is needed if this is not the case. For each point the flow line strip through it is identified and its upper convexity used to give the topographic scale time, t_G. The area drained per unit strip width, a; and the local gradient, g, are also obtained for the point to give the saturated runoff, $j_* \exp(-\gamma)$, at which the point belongs to the contributing area. A curve similar to one of these in Figure 6 can then be drawn from the sample points. In order to represent the hollow areas adequately, it is anticipated that a higher density of sampling points would be required. The two component model outlined above can then give estimates of the slow and quick runoff responses to each storm at slope bases. What the procedure above fails to allow for is differences in timing due to channel flow velocities. The grid basis therefore appears adequate to account for changes in soil moisture due to slope runoff (except along major flood plains), but unsatisfactory for estimating the timing of large stream hydrographs.

A further speculation is that the geomorphology may give additional information to help estimate some other hydrological parameters for each area from remotely sensed data. The channel network density revealed either on a grid or catchment basis reflects the geomorphological evolution of an area in response to its hydrology, and is thought to respond to it over periods of the order of 10^1 to 10^3 years, and so broadly reflect current rather than fossil conditions. Following the stability argument of Smith and Bretherton (1972), Kirkby (1980) has argued that the average slope length $[1/(2\times$ drainage density $)]$ is of the same order as the distance at which overland flow erosion becomes dominant over creep and rainsplash. This distance strongly reflects the rainfall regime, but is also influenced by soil and vegetation. If the rainfall regime is independently known, and vegetation cover can be measured remotely, then hydrological soil factors remain as the unknown to be deduced. Clearly this approach is unlikely to provide all the information required, but may be a worthwhile step towards the assessment of runoff response remotely, for both grid square and catchment units.

REFERENCES

Beven, K.J., 1979. On the Generalized Kinematic Routing Method, *Water Resources Research*, 15: pp. 1238-1242.

Beven, K.J., 1983. Introducing Spatial Variability Into TOPMODEL: Theory and Preliminary Results, *Report to Department of Environmental Sciences*, University of Virginia, 35 pp.

Beven, K.J. and Kirkby, M.J., 1979. A Physically Based Variable Contributing Area Model of Basin Hydrology, *Hydrological Sciences Bulletin*, 24: pp. 43-69.

Beven, K.J., Kirkby, M.J., Schofield, N. and Tagg, A.F., 1984. Testing a Physically-based Flood Forecasting Model (TOPMODEL) for Three U.K. Catchments, *J. Hydrology*, 69: pp. 119-143.

Beven, K.J., and Wood, E.F., 1983. Catchment Geomorphology and the Dynamics of Runoff Contributing Areas, *J. Hydrology*, 65: pp. 139-158.

Carson, M.A., and Kirkby, M.J., 1972. *Hillslope Form and Process*, Cambridge University Press, 475 pp.

Green, W.H., and Ampt, G.A., 1911. Studies on Soil Physics. 1. The Flow of Air and Water Through Soils, *J. Agric. Soils*, 4: pp. 1-24.

Gupta, V.K., Waymire, E., and Rodriguéz-Iturbe, I. 1986. On Scale, Gravity and Network Structure in Basin Runoff, (this issue). D. Reidel Publishing Company.

Kirkby, M.J., 1976. Tests of the Random Network Model, and Its Application to Basin Hydrology, *Earth Surface Processes*, 1: pp. 197-212.

Kirkby, M.J., 1980. The Stream Head as a Significant Geomorphic Threshold, in *Thresholds in Geomorphology*, edited by D.R. Coates, and J.D. Vitek, George Allen & Unwin, 498 pp., Chapter 4, pp. 53-73.

Rodríguez-Iturbe, I., and Valdés, J.B., 1979. The Geomorphologic Structure of Hydrologic Response, *Water Resources Research*, 15: pp. 1490-1420.

Smith, T.R., and Bretherton, F.P., 1972. Stability and the Conservation of Mass in Drainage Basin Evolution, *Water Resources Research*, 8: pp. 1506-1524.

Surkan, A.J., 1969. Synthetic Hydrographs: Effects of Network Geometry, *Water Resources Research*, 5: pp. 112-128.

GEOMORPHOLOGIC APPROACH TO SYNTHESIS OF
DIRECT RUNOFF HYDROGRAPH FROM THE UPPER
TIBER RIVER BASIN, ITALY

4

C. Corradini
F. Melone
L. Ubertini
V.P. Singh

ABSTRACT

The effective rainfall-direct runoff relationship was investigated for forty events on four large basins by using a geomorphologic representation of the instantaneous unit hydrograph (IUH) proposed by Gupta et al. (1980). These basins are a part of the Upper Tiber River basin located in Central Italy and range in area from 934 km^2 to 4,147 km^2. For each event the volume of direct runoff was obtained by hydrograph separation. The effective rainfall hyetograph was then determined by using the two-term Philip infiltration equation in conjunction with a volume balance analysis. The geomorphologic parameters required by the IUH were extracted from a topographic map of each basin with the map scale of 1:200,000. It was found that the dimensionless form of the IUH remained practically constant from one basin to another. By convoluting the geomorphologic IUH, derived for each basin, with the effective rainfall, the direct runoff hydrograph was synthesized for each event. The model results compare reasonably well with observations of each basin. The maximum and mean errors in computing peak flow were 33% and 15% respectively. Furthermore, the magnitudes of these errors did not depend upon the basin area. A sensitivity analysis of the model structure revealed that its order of geomorphologic representation could be reduced by at least one without a significant loss of accuracy. This small reduction in order amounted to a considerable reduction in geomorphologic complexity and computational effort. The model results were quite sensitive to basin lag and sorptivity parameter of the infiltration equation. A 10% variation in basin lag resulted in approximately 10% variation of computed peak discharge. However, a 10% variation in sorptivity produced an approximately 20% mean variation in computed peak discharge.

V. K. Gupta et al. (eds.), Scale Problems in Hydrology, 57–79.
© *1986 by D. Reidel Publishing Company.*

1. INTRODUCTION

In a pioneering study Rodriguéz-Iturbe and Valdés (1979) developed a rather refreshing approach to determination of the instantaneous unit hydrograph (IUH) by explicitly incorporating the characteristics of drainage basin composition (Horton, 1945; Strahler, 1964; Smart, 1972). This approach, designated henceforth as RV approach, coupled the empirical laws of geomorphology with the principles of linear hydrologic systems. Rodriguéz-Iturbe and his associates have since extended this approach by explicitly incorporating climatic characteristics, and have studied several aspects including hydrologic similarity (Rodriguéz-Iturbe et al., 1979; Valdés et al., 1979; Rodriguéz-Iturbe, Gonzalez-Sanabria and Bras, 1982; Rodriguéz-Iturbe, Gonzalez-Sanabria and Caamaño, 1982). Motivated by the work of Rodriguéz-Iturbe and his associates, Gupta et al. (1980) examined the RV approach, reformulated it, simplified it, and made it more general and even more mathematically elegant. Its assumptions, limitations and potential for application to synthesis of runoff from ungaged basins became more apparent. Others (Hebson and Wood, 1982; Kirshen and Bras, 1983; Singh, 1983) have also applied the geomorphologic concepts to hydrologic analyses.

In this study we follow the approach of Gupta et al. (1980), designated henceforth as the GWW approach. These investigators tested their approach by comparing theoretical and observed direct runoff hydrographs for three events that occurred on three basins in Illinois. A good agreement was found for two larger basins (area 2,061 km^2 and 1,658 km^2; basin lag 57.6 hrs and 71.5 hrs) but the peak runoff was underestimated for the smaller basin (area 554 km^2, basin lag 40 hrs). The inadequacy of linearity assumption was indicated as a possible cause of the lack of agreement for this basin (Wang, et al., 1981). However, this approach remains yet to be tested, by using a large number of events, on a wide variety of basins. Furthermore, there are a number of issues that need to be addressed. Although channel network characteristics are explicitly included in the GWW approach, it is not clear as to how much detail is required in representing them. For example, for a 5-order basin is it necessary to consider its 5-order channel network or is it possible to consider a lower order representation? It should be noted that a small reduction in this representation will amount to a large reduction in complexity and computational effort. Further, precision of geomorphologic representation depends on the scale of the topographic map. What scale map should then be used for this representation? Another question relates to sensitivity of this approach to its components and parameters.

The objective of this paper is to address the above issues by applying the GWW approach to four Italian basins, ranging in area from 934 km^2 to 4,147 km^2. Forty rainfall-runoff events were considered. The effective rainfall was determined by using the two-term Philip infiltration equation with a volume balance procedure.

2. THE GEOMORPHOLOGIC APPROACH

If the basin is assumed to be linear and time-invariant then the direct runoff $Q(t)$ at any

$$Q(t) = \int_0^t h(t - \tau)R(\tau)d\tau \qquad (1)$$

where R is the effective rainfall intensity and h the IUH. Since $R(t)$ is assumed to be known, the problem of determining $Q(t)$ reduces to one of determining the IUH. Rodriguéz-Iturbe and Valdés (1979) developed a geomorphologic approach to specify $h(t)$ which was later generalized by Gupta et al. (1980). Although a detailed discussion of this approach can be found in these cited references, we provide here, for the sake of completeness, a short account following Gupta et al. (1980).

A basin of order W contains streams of order from one to W (following Horton-Strahler ordering system). The network of these streams and their drainage areas determine the paths to be followed by rainwater from the point of its landing to the basin outlet. The number of the paths S_f specified by the basin geomorphology will always be less than or equal to 2^{W-1}. A given path is composed of one overland region (r) and one or more channels (c); it can be expressed in terms of states ($x_i, i = 1,2,...,j$) and ($j \leq W + 1$). If we ignore the surface areas of the channels then the area drained by this path is equal to the area of the overland region associated with its lowest order channel. If we define a path $s \in S_f$ of the form $s = \{x_1, x_2, \ldots, x_j\}$, where $\{x_1, x_2, \ldots, x_j\} \in \{r_1, r_2,...,r_W; c_1, c_2, \ldots, c_W\}$, then assuming uniform areal coverage of the effective rainfall, the probability $p(s)$ that this path s, from amongst all possible paths, will be followed can be expressed as

$$p(s) = A_{x_1} P_{x_1, x_2} \cdots P_{x_{j-1}, x_j} \tag{2}$$

where A_{x_1} is the ratio of the overland region area to the basin area, and $P_{x_i, x_{i+1}}$ is the proportion of channels of order i which merge into channels of order $i+1$. Furthermore, the time T_s the water will take to follow this path will be the sum of the times T_{x_i} spent in the various states forming this path,

$$T_s = \sum_{i=1}^{j} T_{x_i}, \quad j > 1 \tag{3}$$

If T_B represents the random time which water spends in the basin then

$$T_B = \sum_{s \in S_f} T_s I_s \tag{4}$$

where I_s is the indicator function and equals 1 if the water follows the path s and 0 otherwise. Let f_{x_i} denote the probability density function (pdf) of T_{x_i} and F_{x_i} its cumulative distribution function (cdf). The cdf of T_B follows from Eqn.(4)

$$P(T_B < t) = \sum_{s \in S_f} P(T_s < t) p(s) \tag{5}$$

$$= \sum_{s \in S_f} [F_{x_1} * F_{x_2} * \cdots * F_{x_j}(t)] p(s)$$

where the asterisk signifies the convolution operation. Differentiating eq. 5 with respect to t,

$$f(T_B) = \sum_{s \in S_f} (f_{x_1} * f_{x_2} * \cdots * f_{x_j}) p(s) \tag{6}$$

It can be shown that the IUH $h(t) = f(T_B)$. This gives an explicit representation of $h(t)$ in terms of basin geomorphology.

Following Rodriguéz-Iturbe and Valdés (1979) and Gupta et al. (1980), f_{x_i} can be assumed to be exponential with some parameter K_{x_i}. Then Eqn. (6) becomes

$$h(t) = \sum_{s \in S_f} \sum_{i=1}^{j} C_{ij} [\exp(-K_{x_i} t)] p(s), \tag{7}$$

where $s = \{x_1, x_2,...,x_j\}$

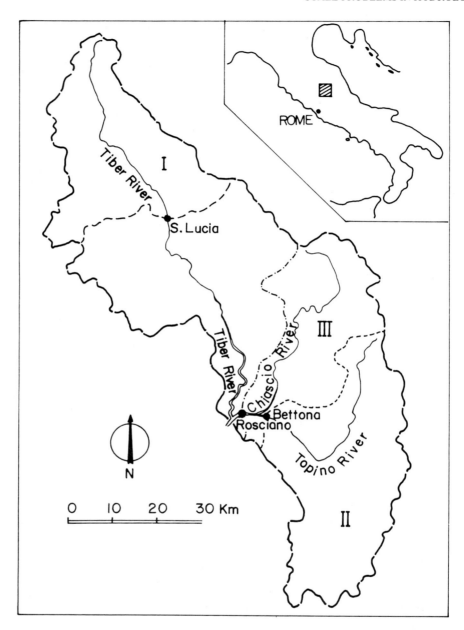

FIGURE 1. The Upper Tiber River basin with its sub-basins: Tiber at S. Lucia (I), Topino at Bettona (II), Chiascio at Rosciano (II + III). The geographic location of the basin is also shown.

The parameter $1/K_{x_i}$ represents the mean holding time of the element x_i. The C_{ij} are coefficients which can be expressed as a function of K_{x_i} by

$$C_{ij} = K_{x_1} K_{x_2} \cdots K_{x_j} [(K_{x_1} - K_{x_i}) \cdots (K_{x_{i-1}} - K_{x_i}) H (K_{x_{i+1}} - K_{x_i}) \cdots (K_{x_j} - K_{x_i})]^{-1} \quad (8)$$

According to Gupta et al. (1980), the mean holding time of an ith order Strahler channel, $x_i \equiv c_i$, is given by

$$\frac{1}{K_{c_i}} = \gamma (\bar{L}_i)^{1/3}, \quad 1 \leq i \leq W \tag{9}$$

where γ is an empirical constant, $\bar{L}_i$ the average channel length of order i. Likewise, the mean holding time of an ith order overland region, $x_i \equiv r_i$, is given by

$$\frac{1}{K_{r_i}} = \gamma \left(\frac{A_{r_i} A}{2 N_i \bar{L}_i} \right)^{1/3}, \quad 1 \leq i \leq W \tag{10}$$

where A_{r_i} is the ratio of the ith order overland region area to the basin area A, N_i is the number of the ith order overland regions. The constant γ in Eqns. (9)-(10) can be derived from

$$K_B = \sum_{s \in S_f} \left[\frac{1}{K_{x_1}} + \frac{1}{K_{x_2}} + \cdots + \frac{1}{K_{x_j}} \right] p(s), \tag{11}$$

$$s = \{x_1, x_2, \ldots, x_j\}$$

where K_B is the mean holding time of the basin or basin lag which can be estimated from basin area (Boyd, 1978; Boyd et al., 1979; Panu and Singh, 1981; Singh, 1983). Thus the IUH is completely specified by basin geomorphology.

3. EXPERIMENTAL BASINS

Four basins were selected for this investigation. These are the Upper Tiber River basin with a drainage area of 4,147 km^2 and three of its sub-basins: Chiascio River at Rosciano with an area of 1,956 km^2, Topino River at Bettona with an area of 1,220 km^2 and Tiber at S. Lucia of 934 km^2. The general layout of these basins and their geographic location are shown in Fig. 1. These basins are characterized by a complex orography. The topography is mainly hilly with elevation above the mean sea level ranging from 200 to 800 m. The mountain peaks on a large portion of the boundary of the Upper Tiber River basin range in elevation from 1,000 to 1,500 m above mean sea level. Probably the assumption of uniform effective rainfall is then questionable.

Mean annual precipitation over the basins ranges from 700 mm to 1,600 mm. Higher monthly precipitation values generally occur during the autumn-winter period (November, December). It is this period during which floods normally occur. The rains causing these floods are widespread. Rainfall infiltration can be linked to the main geologic features of the basins. The Topino River basin is partly (about 1/3 of the basin area) made up of rock formations characterized by strong Karst phenomena. In the remainder of the Upper Tiber River basin the soil, overlaying practically impervious rocks, is made up of clay and sandy silt.

Geomorphologic parameters were estimated by using the map scale 1:200,000 and Horton-Strahler ordering system. They are summarized in Table 1 where it can be seen that Horton's

TABLE 1. Geomorphologic Properties of Upper Tiber River Basin and its Sub-basins

Basin	Stream Order	Area Ratio R_A	Bifurcation Ratio R_B	Stream Length Ratio R_L
Upper Tiber River				
Area 4,147 km ^{2}	1		4.12	
	2	4.23	4.63	2.58
	3	4.90	3.38	2.36
Highest-Order	4	3.49	4.00	2.15
Stream Length	5	6.25	2.00	2.40
126.0 km	6	2.00		1.32
	Average Value	4.17	3.63	2.16
Chiascio River at Rosciano				
Area 1,956 km ^{2}	1		4.39	
	2	4.46	4.08	2.58
	3	4.64	3.00	2.40
Highest-Order	4	3.52	4.00	2.24
Stream Length	5	4.98		1.41
64.6 km	Average Value	4.40	3.87	2.16
Topino River at Bettona				
Area 1,220 km ^{2}	1		4.24	
	2	4.58	3.63	2.74
	3	4.35	2.67	2.41
Highest-order	4	2.61	3.00	1.72
Stream Length	5	4.08		1.57
59.0 km	Average Value	3.91	3.39	2.11
Tiber River at S. Lucia				
Area 934 km ^{2}	1		3.94	
	2	3.71	5.00	2.40
	3	6.83	3.50	2.98
Highest-Order	4	4.11	2.00	1.94
Stream Length	5	2.41		1.47
57.2 km	Average Value	4.27	3.61	2.20

numbers R_A, R_B, R_L range within the limits usually found in natural basins. The parameters R_A and R_L were determined as proposed by Schumm (1956) and Bowden and Wallis (1964), respectively.

4. RAINFALL-RUNOFF DATA

Forty rainfall-runoff events, ten for each basin, were used in this study. Table 2 summarizes these events. There are fifty-four rainfall measuring stations distributed throughout the Upper Tiber River basin; forty-two of these are recording rain gauges and twelve telemetering rain gauges. By using the Thiessen method the point measurements of rainfall were transformed to areal average values for each event. An examination of rainfall data and vertical profile of air temperature indicated that the events selected were not influenced by snow.

Most of the runoff hydrographs are single-peaked. A few of the hydrographs had more than one peak, but were such that a reliable partitioning into single-peak hydrographs was found to be relatively easy. Single-peak hydrographs were preferred because they allowed a better estimation of the effective rainfall for the two-term Philip infiltration equation. For each rainfall-runoff event the direct runoff was obtained by hydrograph separation (Wilson, 1974; Linsley et al., 1982). The procedure used consists of extending the recession (normally at a constant rate) existing before the storm to a point under the peak of the hydrograph, and then connecting by a straight line this point to that of the greatest curvature on the recession limb of the hydrograph.

5. DETERMINATION OF INFILTRATION AND EFFECTIVE RAINFALL

Infiltration for each rainfall-runoff event was determined by using the two-term Philip infiltration equation (Philip, 1969)

$$f(t) = f_\infty + 0.5 \ S \ t^{-0.5} \qquad (12)$$

where $f(t)$, in cm/hr, is the rate of infiltration at time t; f_∞, in cm/hr, a parameter approximately equal to saturated hydraulic conductivity; S, in $cm/hr^{0.5}$, a parameter called sorptivity. For f_∞ a wide range of possible values was determined for each basin on the basis of its main geologic features discussed previously and following Freeze and Cherry (1979, pp. 26-29). However each range was limited by analyzing with S=0 the volume balance of each event. In particular the maximum value for f_∞ was chosen as the largest one producing, for all the events in the basin, a volume of effective rainfall greater than the corresponding volume of direct runoff. The threshold values of 0.08, 0.16, 0.22 and 0.06 cm/hr were computed from the largest basin to the smallest one. In the narrow ranges so obtained reasonable values of f_∞ were selected, as shown in Table 3. Really this procedure is relatively arbitrary, but it is supported by an analysis of model sensitivity to f_∞ which is reported later. The parameter S depends mainly on antecedent soil moisture and soil characteristics. It was determined for each rainfall-runoff event by a volume balance analysis.

This involves choosing a value of S and determining the volume of effective rainfall which is then compared with the volume of direct runoff. That value of S which results in equality of volumes of the effective rainfall and direct runoff is taken as the correct value. The values of S are shown in Table 3. S generally ranges from 0.3 $cm/hr^{0.5}$ to 0.9 $cm/hr^{0.5}$ for the Upper Tiber River basin, 0.4-1.6 $cm/hr^{0.5}$ for the Chiascio River basin at Rosciano, 0.6-1.4 $cm/hr^{0.5}$ for the Topino at Bettona, and 0.2-1.0 $cm/hr^{0.5}$ for the Tiber at S. Lucia. The effective rainfall was determined for each event by subtracting infiltration rate from its rainfall hyetograph. The effective rainfall intensity was zero during periods when the infiltration rate was higher than rainfall intensity. The maximum intensity of effective rainfall did not exhibit a great variability from one event to another for any of the basins considered; its value is generally less than 4.0 mm/hr.

TABLE 2. Some Characteristics of Rainfall-Runoff Events of the Upper Tiber River Basin and its Sub-basins

Serial Number	Date	Rainfall Depth mm	Rainfall Duration hrs	Effective Depth mm	Effective Rainfall Duration hrs	Initial Base Flow m³/s	Time to Peak m³/2	Direct Peak hrs	Runoff Duration hrs
					Upper Tiber River Basin				
1	12/12/81	16.6	12.0	2.9	7.0	60.0	200.0	19.9	48.0
2	12/18/81	19.3	15.0	3.7	8.0	75.0	265.4	19.9	48.0
3	12/24/81	16.3	18.0	2.8	5.5	87.2	139.0	15.4	62.0
4	12/29/81	25.0	18.0	11.7	11.0	83.6	591.0	22.9	70.0
5	11/9/82	49.1	23.0	13.1	9.0	27.0	702.8	18.9	61.0
6	11/13/82	28.2	15.0	6.2	8.0	58.0	404.9	19.9	57.0
7	12/1/82	15.0	7.0	4.9	5.4	48.0	338.0	20.3	54.0
8	12/13/82	18.0	13.0	3.2	10.0	135.0	188.7	17.9	56.0
9	12/18/82	32.5	11.0	11.9	7.0	83.0	760.3	14.9	53.0
10	12/21/82	50.4	45.0	17.8	27.0	105.6	629.0	26.9	83.0
					Chiascio River Basin at Rosciano				
1	12/11/81	27.7	12.0	1.3	4.0	9.5	45.5	13.9	44.0
2	12/12/81	15.4	7.0	1.5	3.0	12.5	58.5	9.9	43.0
3	12/13/81	16.7	10.0	1.8	3.0	24.0	67.0	8.9	47.0
4	12/29/81	19.7	18.0	3.4	4.0	36.3	107.7	17.9	63.0
5	11/9/82	47.0	12.0	5.3	1.0	9.7	165.3	15.9	51.0
6	11/18/82	26.5	10.0	2.8	4.0	19.0	116.7	14.9	48.0
7	12/1/82	12.6	7.0	2.5	4.0	12.5	83.5	16.9	45.0
8	12/13/82	25.9	13.0	3.2	9.0	45.0	87.3	14.9	50.0
9	12/18/82	42.1	11.0	10.3	7.0	37.5	350.0	14.9	48.0
10	12/21/82	72.9	43.0	17.6	25.0	52.7	339.8	25.9	69.0

TABLE 2. (cont.) Some Characteristics of Rainfall-Runoff Events of the Upper Tiber River Basin and its Sub-basins

Serial Number	Date	Rainfall Depth mm	Rainfall Duration hrs	Effective Rainfall Depth mm	Rainfall Duration hrs	Initial Base Flow m³/s	Time to Peak m³/2	Direct Runoff Peak hrs	Direct Runoff Duration hrs
				Topino River Basin at Bettona					
1	12/12/81	30.4	12.0	1.2	5.0	12.2	33.6	9.9	44.0
2	12/12/81	16.6	7.0	1.4	3.0	14.0	42.0	8.9	44.0
3	12/13/81	17.6	10.0	1.3	1.0	18.3	36.0	7.9	42.0
4	12/24/81	16.4	16.0	1.4	2.0	23.9	28.8	7.9	45.0
5	12/29/81	20.3	18.0	1.7	3.9	20.4	32.8	12.8	54.0
6	11/9/82	39.6	11.0	1.4	1.0	6.0	27.0	13.9	47.0
7	11/13/82	30.4	10.0	2.0	4.9	8.0	56.3	14.8	46.0
8	12/13/82	26.5	13.0	2.7	8.0	11.9	50.0	14.9	40.0
9	12/18/82	39.7	11.0	6.0	6.5	13.5	159.6	12.4	42.0
10	12/21/82	70.1	31.0	13.9	23.0	19.2	186.2	21.9	65.0
				Tiber River at S. Lucia					
1	12/11/81	12.8	10.0	2.5	4.0	8.5	58.4	10.9	34.0
2	12/12/81	19.2	11.0	7.4	9.2	10.5	163.0	16.1	41.0
3	12/18/81	23.6	19.0	10.3	11.0	19.5	192.0	13.9	37.0
4	12/19/81	11.8	6.0	5.6	5.1	22.5	120.0	9.0	35.0
5	11/9/82	36.4	20.0	13.4	9.0	4.0	220.8	10.9	48.0
6	11/13/82	36.4	14.0	13.5	11.0	15.0	256.0	16.9	37.0
7	12/1/82	16.8	10.0	12.0	8.6	21.0	243.9	12.5	33.0
8	12/18/82	14.3	9.0	4.3	3.0	25.2	78.3	8.9	39.0
9	12/21/82	7.6	5.0	1.7	3.4	29.0	46.2	8.3	31.0
10	12/22/82	11.4	15.0	3.2	13.2	33.0	41.0	11.1	45.0

TABLE 3. Parameters f_∞ and S of the Philip Infiltration Equation

Event Serial Number	Upper Tiber River Basin	Chiascio River Basin at Rosciano	Topino River Basin at Bettona	Tiber River Basin at S. Lucia
		S, cm/hr $^{0.5}$		
1	0.511	1.341	1.151	0.751
2	0.511	0.771	0.631	0.391
3	0.421	0.851	1.011	0.421
4	0.321	0.901	1.141	0.281
5	1.761	3.031	0.971	0.971
6	0.881	1.251	4.761	0.841
7	0.361	0.411	1.281	0.141
8	0.451	0.701	0.571	0.761
9	0.921	1.561	1.421	0.341
10	0.491	1.051	0.831	0.211
		f_∞, cm/hr		
		0.030	0.050	0.100

6. DETERMINATION OF BASIN LAG

The basin lag in Eqn. (11) was determined by averaging the observed lags for all the events on each basin. Table 4 presents values of K_B together with the standard deviation. These values appear to follow the relation

$$K_B = b \ A^\alpha \tag{13}$$

where K_B is in hrs and A basin area in km^2. For $\alpha = 0.38$ (Boyd, 1978; Boyd et al., 1979; Panu and Singh, 1981; Singh, 1983) an optimal value of 0.823 for b was found for all four basins. Errors of less than 10% would be expected in K_B from Eqn. (13) for these basins.

TABLE 4. Basin lag times

Basin	Lag Time	Mean Deviation hrs
Upper Tiber River	18.0	2.824
Chiascio River at Rosciano	14.7	1.935
Topino River at Bettona	13.3	1.973
Tiber River at S. Lucia	11.5	1.565

7. DIRECT RUNOFF COMPUTATION

7.1. Determination of the instantaneous unit hydrograph

Using the parameters estimated in the above manner the IUH was determined for each basin, as shown in Fig. 2(A). It can be seen that the IUH possesses appropriate shape characteristics. Because the basins are large, the IUH experiences a slow rise and a slow recession. As the basin area increases, lower rates of rise and recession are observed. Furthermore the IUH peak appears to be linked to basin area. Fig. 2(B) presents the IUH in dimensionless form which was determined by dividing the ordinate by peak value and abscissa by basin lag. It is practically the same for all the basins; the IUH corresponding to the largest basin shows only a slight deviation. This then implies that a dimensionless IUH can be prescribed for these basins and the IUH for any of these basins can be determined from this and one normalizing quantity, either basin lag or IUH peak.

7.2. Comparison of computed and observed direct runoff

The direct runoff hydrographs were computed for each event on each basin by using Eqn. (1). A six-minute discretization time interval was used for the computations. A comparison of computed and observed hydrograph characteristics is shown in Figs. 3-6 for one sample event on each of the basins of the Upper Tiber River, Chiascio River at Rosciano, Topino River at Bettona, Tiber River at S. Lucia, respectively. Furthermore, Figs. 7-10 summarize observed and computed peak runoff values, and Figs. 11-14 compare observed and computed values of time to peak runoff. The mean characteristics of computed and observed hydrographs are generally in reasonable agreement as exhibited by the sample events in Figs. 3-6. For the Upper Tiber River basin the maximum peak runoff error ϵ_{Q_P} is 27% and its mean value $\epsilon_{Q_P^m}$ is 16%; the corresponding values for the time to peak are $\epsilon_{t_P} = 17\%$ and $\epsilon_{t_P^m} = 8\%$. By combining Figs. 7 and 11 and dividing the errors in four groups, it follows that the computed peak flow is anticipated and underestimated in $n_1=4$ cases; anticipated and overestimated in $n_2=2$ cases; delayed and underestimated in $n_3=2$ cases; delayed and overestimated in the other cases, $n_4=2$. For the Chiascio River basin at Rosciano, ϵ_{Q_P}, $\epsilon_{Q_P^m}$, ϵ_{t_P} and $\epsilon_{t_P^m}$ are 28%, 13%, 29% and 18% respectively. As to the four groups considered, from Figs. 8 and 12 it follows that $n_1=6$, $n_2=1$, $n_3=2$, $n_4=1$. For the Topino River basin at Bettona, $\epsilon_{Q_P}=31\%$, $\epsilon_{Q_P^m}=16\%$, $\epsilon_{t_P}=32\%$, $\epsilon_{t_P^m}=18\%$ and from Figs. 9 and 13, $n_1=1$, $n_2=3$, $n_3=4$, $n_4=2$. For the Tiber River basin at S. Lucia, $\epsilon_{Q_P}=33\%$, $\epsilon_{Q_P^m}=15\%$, $\epsilon_{t_P}=25\%$, $\epsilon_{t_P^m}=13\%$ and, from Figs. 10 and 14, it follows $n_1=4$, $n_2=0$, $n_3=4$, $n_4=2$. The results obtained show that the model accuracy is practically the same for the four basins analyzed so no relationship between model accuracy and basin area is apparent. On the other hand, as it can be seen from Figs. 7-14, the goodness of the theoretical results is weakly linked to peak runoff. In fact, the hydrographs with the largest values of peak runoff, to which the largest values of effective rainfall intensity are generally associated, are computed a little better than those with the smallest values of peak runoff.

7.3. Model sensitivity analysis

An analysis was made of the sensitivity of model results to changes in the effective rainfall, basin lag K_B and geomorphologic representation of the basin according to Horton-Strahler ordering scheme.

Errors in the hydrograph separation procedure can lead to errors ϵ_V in computation of the effective rainfall volume. It was found that the computed hydrograph was significantly

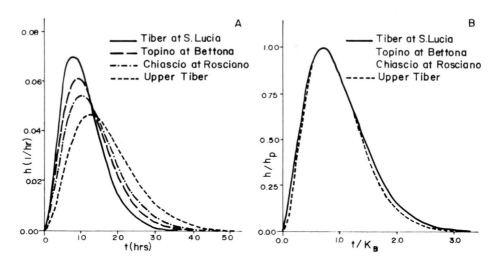

FIGURE 2. (A) The instantaneous unit hydrographs (IUH) of the four basins indicated. (B) The dimensionless IUH's. K_B is the basin lag, h_P is the IUH peak value.

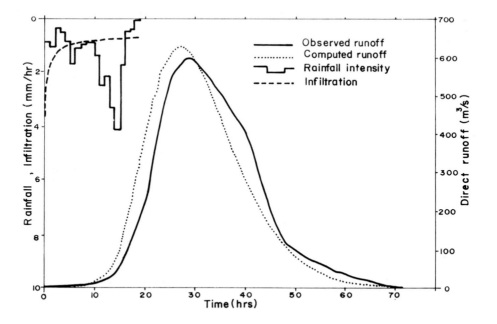

FIGURE 3. Comparison of observed and computed direct runoff hydrographs for the event of December 29, 1981, on the Upper Tiber River basin.

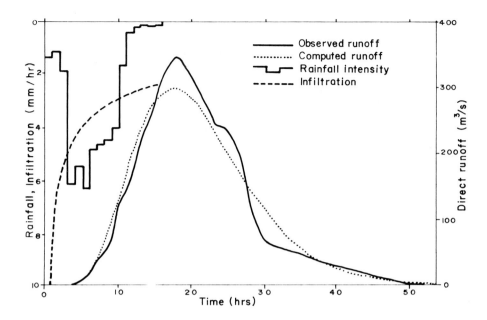

FIGURE 4. Comparison of observed and computed direct runoff hydrographs for the event of December 18, 1982, on the Chiascio River basin at Rosciano.

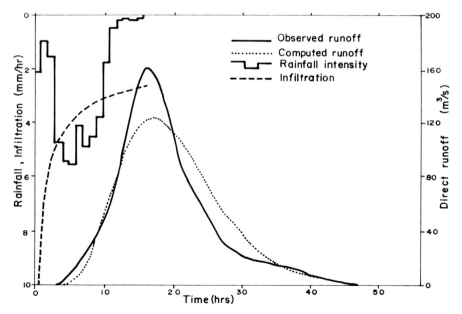

FIGURE 5. Comparison of observed and computed direct runoff hydrographs for the event of December 18, 1982, on the Topino River basin at Bettona.

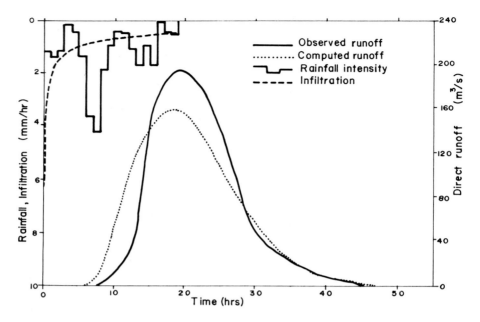

FIGURE 6. Comparison of observed and computed direct runoff hydrographs for the event of December 18, 1981, on the Tiber River basin at S. Lucia.

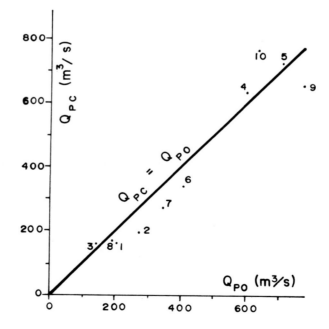

FIGURE 7. Relationship between computed and observed direct runoff peaks, Q_{PC} and Q_{PO} respectively, for all the events analyzed for the Upper Tiber River basin.

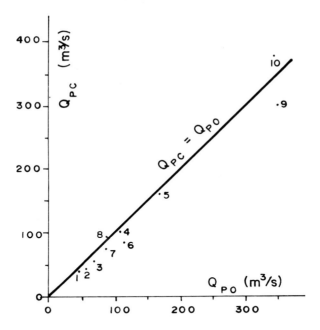

FIGURE 8. Relationship between computed and observed direct runoff peaks, Q_{PC} and Q_{PO} respectively, for all the events analyzed for the Chiascio River basin at Rosciano.

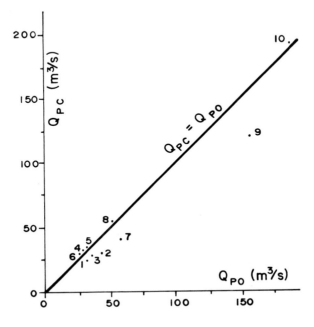

FIGURE 9. Relationship between computed and observed direct runoff peaks, Q_{PC} and Q_{PO} respectively, for all the events analyzed for the Topino River basin at Bettona.

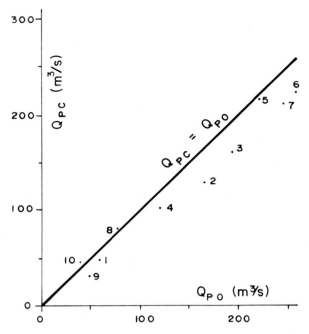

FIGURE 10. Relationship between computed and observed direct runoff peaks, Q_{PC} and Q_{PO} respectively, for all the events analyzed for the Tiber River basin at S. Lucia.

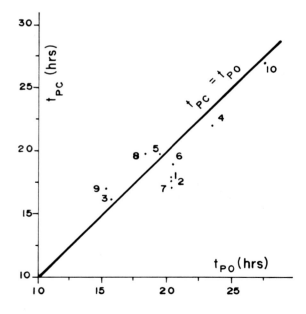

FIGURE 11. Relationship between computed and observed times to peak, t_{PC} and t_{PO} respectively, for all the events analyzed for the Upper Tiber River basin.

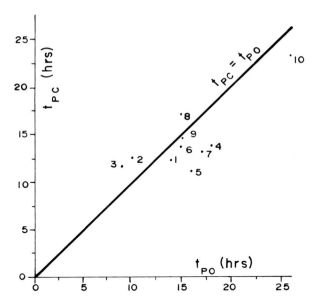

FIGURE 12. Relationship between computed and observed times to peak, t_{PC} and t_{PO} respectively, for all the events analyzed for the Chiascio River basin at Rosciano.

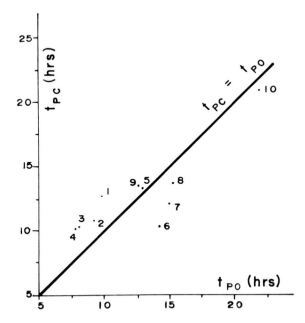

FIGURE 13. Relationship between computed and observed times to peak, t_{PC} and t_{PO} respectively, for all the events analyzed for the Topino River basin at Bettona.

TABLE 5. Mean Values of the Percent Variations in Computed Peak of Direct Runoff and Time to Peak $\overline{\Delta Q}_{PC}$ and $\overline{\Delta t}_{PC}$ respectively, for Various Errors in Effective Rainfall Volume

	-20%		-10%		+10%		+20%	
Basin	$\overline{\Delta Q}_{PC}$	$\overline{\Delta t}_{PC}$	$\overline{\Delta Q}_{PC}$	$\overline{\Delta t}_{PC}$	$\overline{\Delta Q}_{PC}$	$\overline{\Delta t}_{PC}$	$\overline{\Delta Q}_{PC}$	$\overline{\Delta t}_{PC}$
Upper Tiber River	19.5	3.2	9.3	1.1	10.2	4.7	19.9	6.6
Chiascio River at Rosciano	19.9	1.0	9.8	0.7	10.1	1.3	20.0	2.2
Topino River at Bettona	19.6	3.4	10.0	2.6	9.9	0.4	20.1	1.1
Tiber River at S. Lucia	19.7	2.4	9.5	0.8	10.1	3.8	20.5	5.8

The percent variation for a quantity was computed as the difference in the quantity X computed with supposedly correct model parameters and that Y computed with erroneous model parameters divided by 0.01X, ignoring algebraic sign. This definition is employed throughout the text. Furthermore for each hydrograph a different time evolution of base flow, represented by a straight line connected the point of rise to that of the greatest curvature on the recession limb, was considered. This procedure generally produced variations of a few units in direct runoff peak errors and virtually no variations in time to peak errors.

Errors in model results produced by an incorrect estimation of the parameter f_∞, but keeping the volume balance condition, were also investigated. It was found that computed direct runoff peak and time to peak were practically unaffected by variations of f_∞ inside the uncertainty range discussed previously. The sensitivity of the model to S of the Philip infiltration equation is of interest when the volume balance procedure is removed. Table 6 lists, for each basin, extreme values of ΔQ_{PC} and Δt_{PC} for various variations of S. As it can be seen a 10% variation of S produced ΔQ_{PC} and Δt_{PC} values up to 91% and 22%, but their mean values were 21% and 2%, respectively.

The computed direct runoff hydrograph was found to be quite sensitive to K_B. In fact a variation of 10% in K_B produced an approximate variation of the same order in ΔQ_{PC} and Δt_{PC}. This is shown in Table 7, where the results for 10% and 20% variations in K_B are reported.

Because a large scale map (1:200,000) was used to represent the geomorphologic parameters, the reliability of this representation was analyzed by varying the basin order W. For each basin the IUH was successfully computed by reducing its order. A comparison of the IUH's associated with the orders W-j (j = 0,1,2,3) was carried out. Rather similar results were obtained for the four basins. As it can be seen from Fig. 15 corresponding to the Topino River basin, only small differences become pronounced for the lower orders. The time to peak decreases by about a quarter of hour as the basin order decreases from W to W-1, one hour from W to W-2, two hours from W to W-3. Therefore, it is reasonable to argue that the scale of the map used here is acceptable. Furthermore, a basin of order W could be reduced to one of order W-1 without causing appreciable error. Table 8 lists, for each basin, the mean values of the percent changes in computed direct runoff peak and its time with basin order representation.

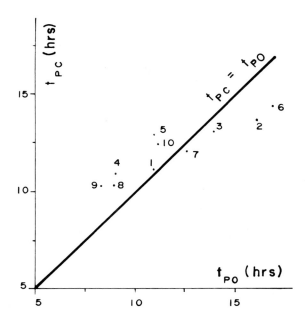

FIGURE 14. Relationship between computed and observed times to peak, t_{PC} and t_{PO} respectively, for all the events analyzed for the Tiber River basin at S. Lucia.

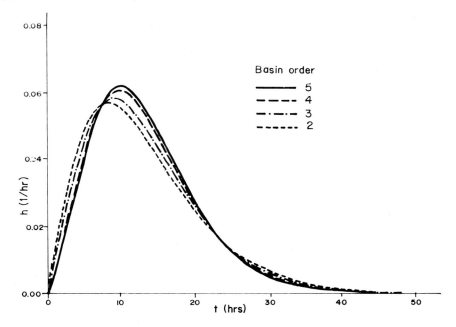

FIGURE 15. The instantaneous unit hydrographs derived by successively reducing the order of the Topino River basin at Bettona.

TABLE 6. Extreme Values of the Percent Changes, ΔQ_{PC} and Δt_{PC}, in Computed Peak of Direct Runoff and Its Time for Various Variations of Sorptivity

Basin	\-20% ΔQ_{PC}	Δt_{PC}	\-10% ΔQ_{PC}	Δt_{PC}	+10% ΔQ_{PC}	Δt_{PC}	+20% ΔQ_{PC}	Δt_{PC}
Upper Tiber River	8.9 -	0.0 -	4.4 -	0.0 -	3.8 -	0.0 -	7.2 -	0.0
	48.2	41.2	23.1	3.7	22.5	8.2	43.5	14.8
Chiascio River at Rosciano	26.3 -	0.0 -	13.0 -	0.0 -	11.3 -	0.0 -	22.1 -	0.0
	195.5	8.8	91.1	8.7	56.3	3.9	90.3	4.7
Topino River at Bettona	29.6 -	0.0 -	13.6 -	0.0 -	9.0 -	0.0 -	16.5 -	0.0
	188.6	8.3	86.0	3.0	73.8	21.8	94.2	21.8
Tiber River at S. Lucia	5.4 -	0.7 -	2.7 -	0.0 -	2.5 -	0.0 -	4.9 -	0.0
	44.0	26.7	20.8	6.3	18.3	2.9	34.9	6.4

TABLE 7. Mean Values of the Percent Changes in Computed Peak of Direct Runoff and Time to Peak, $\overline{\Delta Q}_{PC}$ and $\overline{\Delta t}_{PC}$ respectively, for Various Variations of Basin Lag Time

Basin	\-20% $\overline{\Delta Q}_{PC}$	$\overline{\Delta t}_{PC}$	\-10% $\overline{\Delta Q}_{PC}$	$\overline{\Delta t}_{PC}$	+10% $\overline{\Delta Q}_{PC}$	$\overline{\Delta t}_{PC}$	+20% $\overline{\Delta Q}_{PC}$	$\overline{\Delta t}_{PC}$
Upper Tiber River	21.8	14.0	9.9	7.0	8.3	7.1	15.4	14.3
Chiascio River at Rosciano	22.1	15.5	10.0	7.9	8.3	7.9	15.4	15.8
Topino River at Bettona	22.3	15.7	10.1	7.4	8.4	8.0	15.4	16.3
Tiber River at S. Lucia	20.3	15.2	9.3	7.7	7.7	7.8	14.4	14.9

8. CONCLUSIONS

1.   The linear geomorphologic IUH representation by Gupta et al. (1980), with the two-term Philip infiltration equation, seems to be a reasonable approach to synthesize the direct runoff from large basins.

TABLE 8. **Mean Values of the Percent Changes in Computed Peak of Direct Runoff and Time to Peak, $\overline{\Delta Q}_{PC}$ and $\overline{\Delta t}_{PC}$ respectively, for Different Basin Order Reductions**

Basin	W	W - 1		W - 2		W - 3	
		$\overline{\Delta Q}_{PC}$	$\overline{\Delta t}_{PC}$	$\overline{\Delta Q}_{PC}$	$\overline{\Delta t}_{PC}$	$\overline{\Delta Q}_{PC}$	$\overline{\Delta t}_{PC}$
Upper Tiber River	6	1.7	1.7	4.4	4.8	4.8	8.1
Chiascio River at Rosciano	5	2.0	2.6	5.5	6.9	6.5	12.3
Topino River at Bettona	5	1.7	2.5	5.9	7.2	7.4	14.1
Tiber River at S. Lucia	5	2.4	2.7	4.8	5.5	6.5	10.0

2. The reliability of this representation is not necessarily linked to the basin area. This result is consistent with the observation, often found in the literature, that nonlinear effects in the effective rainfall-direct runoff transformation are not much pronounced in "large" basins.

3. For large basins, geomorphologic parameters can be reliably estimated by using a map scale 1:200,000. Moreover, the basin order so derived can be reduced by one unit without a significant loss of accuracy in the IUH. This small reduction in basin order representation results in a large reduction of geomorphologic complexity and computational effort.

4. The dimensionless IUH is practically the same for all four basins analyzed. This suggests that it may be plausible to specify at least one dimensionless IUH for a number of basins grouped according to some criterion such as geomorphologic similarity.

5. The determination of the infiltration is crucial to direct runoff synthesis. The model results are found to be very sensitive to the S parameter of the Philip infiltration equation used here.

The errors in the model can be due to a variety of causes: errors in effective rainfall, basin lag, representation of mean holding time, etc. Introduction of spatial distribution of rainfall as well as the use of more recent infiltration equations should be explored. Other geomorphologic approaches have recently been proposed which employ channel response forms different from the exponential one used here (Kirshen and Bras, 1983), or incorporate a network geometry different from the Strahler ordering procedure (Gupta and Waymire, 1983). The accuracy of these approaches should be verified on real world basins and then be compared with the one used here.

ACKNOWLEDGMENTS

This research was mainly financed by the National Research Council of Italy under the Bilateral Project Italy-USA and the Special Project IPRA. Furthermore, the participation of V.P. Singh was supported in part by National Science Foundation under the Project, Stochastic Modeling of Streamflow with a Physical Basis, No. NSF-CEE 79-07793. The authors wish to thank B. Bani, C. Fastelli and R. Rosi for their assistance in data acquisition.

REFERENCES

Bowden, K.L. and Wallis, J.R., 1964. Effect of stream ordering technique on Horton's laws of drainage composition, *Geol. Soc. Amer. Bull.*, 75, 767-774.

Boyd, M.J., 1978. A storage-routing model relating drainage basin hydrology and geomorphology, *Water Resources Research*, 14(5), 1921-1928.

Boyd, M.J., Pilgrim, D.H. and Cordery, I., 1979. A storage routing model based on catchment geomorphology, *J. Hydrology*, 42, 209-230.

Freeze, R.A. and Cherry, J.A., 1979. *Groundwater*, Prentice Hall, Englewood Cliffs, NJ, 588 pp.

Gupta, V.K., Waymire, E. and Wang, C.T., 1980. A representation of an instantaneous unit hydrograph from geomorphology, *Water Resources Research*, 16(5), 855-862.

Gupta, V.K. and Waymire, E., 1983. On the formulation of an analytical approach to hydrologic response and similarity at the basin scale, *J. Hydrology*, 65, 95-123.

Hebson, C. and Wood, E.F., 1982. A derived flood frequency distribution using Horton order ratios, *Water Resources Research*, 18(5), 1509-1518.

Horton, R.E., 1945. Erosional development of streams and their drainage basin: hydrophysical approach to quantitative morphology, *Geol. Soc. Amer. Bull.*, 56, 275-370.

Kirshen, D.M. and Bras, R.L., 1983. The linear channel and its effect on the geomorphologic IUH. *J. Hydrology*, 65, 175-208.

Linsley, R.K., Kohler, M.A. and Paulhus, J.L.H., 1982, *Hydrology for Engineers*, McGraw Hill, NY, 508 pp.

Panu, U.S. and Singh, V.P., 1981. Basin lag. Tech. Rep. MSSU-EIRS-CE-84-4, Engineering and Industrial Research Station, Mississippi State University, Mississippi State, MS, 64 pp.

Philip, J.R., 1969. Theory of infiltration, *Advan. Hydrosci.*, 5, 216-296.

Rodríguez-Iturbe, I. and Valdés, J.B., 1979. The geomorphologic structure of hydrologic response. *Water Resources Research*, 15(6), 1409-1420.

Rodríguez-Iturbe, I., Devoto, G. and Valdés, J.B., 1979. Discharge response analysis and hydrologic similarity: the interrelation between the geomorphologic IUH and the storm characteristics. *Water Resources Research*, 15(6), 1435-1444.

Rodríguez-Iturbe, I., Gonzalez-Sanabria, M. and Bras, R.L., 1982. A geomorphoclimatic theory of the instantaneous unit hydrograph, *Water Resources Research*, 18(4), 877-886.

Rodríguez-Iturbe, I., Gonzalez-Sanabria, M. and Caamaño, G., 1982. On the climatic dependence of the IUH: a rainfall-runoff analysis of the Nash model and the geomorphoclimatic theory. *Water Resources Research*, 18(4), 887-903.

Schumm, S.A., 1956. Evolution of drainage systems and slopes in badlands at Perth Amboy, New Jersey. *Geol. Soc. Amer. Bull.*, 67, 597-646.

Singh, V.P., 1983. A geomorphic approach to hydrograph synthesis, with potential for application to ungaged watersheds. Tech. Rep., Water Resources Research Institute, Louisiana State University, Baton Rouge, LA, 101 pp.

Smart, J.S., 1972. Channel networks. *Advan. Hydrosci.*, 8, 305-346.

Strahler, A.N., 1964. Quantitative geomorphology of drainage basins and channel networks. In V.T. Chow (Ed.), *Handbook of Applied Hydrology*, McGraw Hill Book Co., NY, sect. 4-II.

Valdés, J.B., Fiallo, Y. and Rodríguez-Iturbe, I., 1979. A rainfall-runoff analysis of the geomorphologic IUH. *Water Resources Research*, 15(6), 1421-1434.

Wang, C.T., Gupta, V.K. and Waymire, E., 1981. A geomorphologic synthesis of nonlinearity in surface runoff. *Water Resources Research*, 17(3), 545-554.

Wilson, E.M., 1974. *Engineering Hydrology*, McMillan Press, London, 232 pp.

5

SPATIAL HETEROGENEITY AND SCALE IN THE
INFILTRATION RESPONSE OF CATCHMENTS

M. Sivapalan
E.F. Wood

ABSTRACT

The effect of spatial heterogeneity in soil and rainfall characteristics on the infiltration response of catchments is studied. Quasi-analytical expressions are derived for the statistics of the ponding time and the infiltration rate for two cases: (i) spatially variable soils and uniform rainfall, and (ii) constant soil properties and spatially variable rainfall. The derivations show that the cumulative ponding time distribution is a critical variable which governs the mean and covariance of the infiltration process. This distribution determines the proportion of the catchment which is soil controlled and the proportion which is rainfall controlled. The heterogeneity of the infiltration response, part being rainfall controlled and part soil controlled, causes a temporal variation in the correlograms. Over time, the correlation of the infiltration goes from the correlogram of the rainfall (at initial time) to that of the soil properties (at large time).

1. INTRODUCTION

The influence of spatial heterogeneity on hydrologic response and its importance in rainfall runoff modeling has been recognized by hydrologists in recent years. It has been observed empirically that the form of the hydrologic response changes, becoming in general more simple and more linear as the size of the catchment increases with respect to the spatial scale of the heterogeneities (Dooge 1981). Spatial heterogeneity in catchment response arises from the spatial variabilities in climatic inputs such as rainfall, in soil properties such as the soil hydraulic conductivity and in topography. This paper is the first part of a comprehensive research effort on scale problems in runoff production in catchments.

In this paper we are concerned with understanding the influence of rainfall and soil variability on runoff production. In particular we wish to analyze the relative roles these sources of variability have in modeling catchments. In doing so, we hope to answer two questions: (1) Is

81

V. K. Gupta et al. (eds.), Scale Problems in Hydrology, 81–106.
© *1986 by D. Reidel Publishing Company.*

rainfall or soil variability more important in modeling rainfall-runoff processes? (2) Are there scales for which the variability plays a minor role?

To answer these questions, we analyzed two situations: these are the infiltration responses due to (i) spatially variable soil hydraulic conductivity and uniform rainfall and topography and (ii) spatially variable rainfall and uniform soil properties and topography. The more general cases, where all three variables are spatially variable, are analyzed in a later paper (Wood et al., 1986).

The effects of spatial heterogeneity in soil properties on the infiltration response have been studied before, generally by Monte Carlo simulation techniques. Some examples are the work by Sharma and Seely (1979) and Smith and Hebbert (1979) which studied the effect of spatial variability of soil properties on areal rainfall infiltration. Monte Carlo simulation methods do not in general give much insight into the underlying averaging that takes place in nature, or to possible parameterizations of areal infiltration response. Alternative analytical solutions, whenever possible, would be far more preferable.

Maller and Sharma (1981, 1984) presented a direct mathematical procedure to obtain the distributions of the time to ponding, infiltration rate, and cumulative infiltration volume from infiltration parameters of known variability. Their results corroborated the findings of Sharma and Seely (1979) and of Smith and Hebbert (1979). A similar approach was adopted by Milly and Eagleson (1982) who obtained the areal average infiltration due to spatially variable soil properties and rainfall. The fundamental weakness of all this prior work is the failure to incorporate the correlation structure of the soil. By ignoring the soil property correlation, the implicit assumption is made that large areas are being analyzed. Therefore, the research in the literature, constant soil properties or random (but independent) soil properties correspond to very small areas or very large areas, respectively. Little insight is gained into the transition between these extremes and the question of catchment scale cannot be addressed.

In this paper, we propose a methodology that in the presence of spatial heterogeneities in soil properties and/or rainfall enables us to determine quantities such as the areal mean, variance, and autocovariance of the infiltration rates, as well as the probability distributions for ponding time and infiltration rate in the form of quasi-analytical expressions. This analysis will then allow us to address the two questions posed earlier in this section: the relative influence of soil and rainfall variability and the influence of scale on variability effects.

2. MODEL OF RAINFALL INFILTRATION

For completeness and to emphasize a few points in the parameterization of the rainfall-infiltration process, we include a brief but general mathematical statement of the problem being considered. We consider water flow in a soil that lies in the infinite half space beneath the horizontal plane (x, y) with z, the vertical coordinate, directed downward. We assume at the outset, for simplicity, that the soil properties are uniform in the vertical but vary randomly in the plane. In addition, it is assumed that the flow is purely vertical in each profile.

Under these circumstances, the governing equation of soil water movement in each vertical

$$\frac{\partial \theta}{\partial t} + \frac{\partial}{\partial z} \left(K \frac{d\psi}{d\theta} \frac{\partial \theta}{\partial z} \right) + \frac{\partial K}{\partial z} = 0 \qquad (1)$$

where θ is the moisture content, ψ is the suction head, and K is the hydraulic conductivity. Following Dagan and Bresler (1983), the empirical constitutive relations $\psi(\theta)$ and $K(\psi)$ are taken for simplicity as

$$\frac{K(\psi)}{K_s} = \left(\frac{\psi_w}{\psi}\right)^\eta, \quad \psi > \psi_w \tag{2a}$$

$$K(\psi) = K_s, \quad \psi < \psi_w \tag{2b}$$

$$s = \frac{\theta - \theta_r}{\theta_s - \theta_r} = \left(\frac{\psi_w}{\psi}\right)^\beta, \quad \psi > \psi_w \tag{3a}$$

$$s = 1, \qquad \psi < \psi_w \tag{3b}$$

where, K_s is the saturated hydraulic conductivity. ψ_w is the air entry value of suction, s is the reduced water content, θ_r and θ_s are the residual and saturation water contents respectively, and η and β are constants. In the present case, given any one of the soil hydraulic properties ψ, θ, and K, and the six constants, θ_s, θ_r, ψ_w, K_s, η, and β, we can completely characterize the remaining two properties.

We consider infiltration due to rainfall that is constant in time. In such a case, the boundary and initial conditions are as follows:

$$\theta = \theta_i \qquad t = 0 \qquad z > 0 \tag{4}$$

$$q = p \qquad t > 0 \text{ for } \theta(0,t) < \theta_s \tag{5}$$

where $\theta_i > \theta_r$ is the constant initial moisture content, q is the vertical soil water flux and p is the rainfall rate. The ponding time t_p is defined as the time where $\theta(0,t)$ reaches θ_s. Then for $t > t_p$, (5) is replaced by

$$\theta = \theta_s, \qquad z = 0, \qquad t \geq t_p \tag{6}$$

For simplicity, we also neglect the effect of water accumulation at the surface and the effects of surface water "run-on" due to the accumulated water up-gradient running on to neighboring areas and contributing to the infiltration and/or accumulation there.

The purpose of this study is to solve (1) given the constitutive relations (2) and (3) and the boundary and initial conditions (4) to (6) in order to obtain the surface infiltration rates at all points for the case where the six parameters θ_s, θ_r, ψ_w, K_s, η, and β vary randomly in space. However, once again for simplicity, we assume that only K_s is spatially variable while the other five, namely, θ_r, θ_s, ψ_w, η, and β are constant. Dagan and Bresler (1983) have justified this assumption by stating that K_s is much more variable and its variability has a significantly larger impact than variability in the remaining variables.

The solution of Richard's equation (1), even for a fixed K_s, is a formidable task in itself, whose difficulty increases when one desires solutions for areal average infiltration in spatially variable soils. However, Dagan and Bresler (1983) have demonstrated that it is possible to obtain quite accurate estimates of expectations and variances for variable soils by using simpler, approximate solutions to Equation (1). Their arguments are based on the assumption that mutual cancellation of errors occur due to inaccurate flow models, the level of variability in the soil properties and to measurement uncertainties. Dagan and Bresler used a Green-Ampt type infiltration model in their work. Here we use an approximate model for point rainfall infiltration based on the Philip (1957) equation combined with the time compression approximation of Sherman (1943) and Reeves and Miller (1975). The parameters of this model can also be easily and accurately related to the six parameters in the constitutive relations (2) and (3).

The simplified Philip equation is as follows:

$$g^* = \frac{1}{2} S \ t^{-1/2} + cK_s \tag{7}$$

where g^* is the infiltration capacity or infiltration due to instant ponding at time $t = 0$, S is the sorptivity, and K_s is the saturated hydraulic conductivity. S has to be either measured in the field or can be obtained from the constitutive relations (2) and (3). Brutsaert (1976) obtained an expression for S of the following form:

$$S = \left[2(\theta_s - \theta_i) \int_0^1 s^{1-\epsilon} D(s) ds \right]^{1/2} \tag{8}$$

where s is the reduced water content, D is a normalized soil water diffusivity, and ϵ is a weighting parameter which is an independent soil property. For most soils, Brutsaert suggested a value of $\epsilon = 0.5$.

Using the constitutive relations (2) and (3) in (8), we obtained the following approximate relation for S:

$$S = \left\{ \frac{2K_s (\theta_s - \theta_i)^2 \psi_w}{(\theta_s - \theta_r)} \left[\frac{1}{(\eta + 0.5\beta - 1)} + \frac{\theta_s - \theta_r}{\theta_s - \theta_i} \right] \right\}^{1/2} \tag{9}$$

Note that S is proportional to $K_s^{1/2}$ in addition to being dependent on θ_r, θ_s, ψ_w, η, β, and θ_i. Thus, Equation (9) can be rewritten as

$$S = S_r \ K_s^{1/2} \tag{10}$$

where

$$S_r = \left\{ \frac{2(\theta_s - \theta_i)^2 \psi_w}{(\theta_s - \theta_r)} \left[\frac{1}{(\eta + 0.5\beta - 1)} + \frac{\theta_s - \theta_r}{\theta_s - \theta_i} \right] \right\}^{1/2} \tag{11}$$

It can be noted here that the relationship given in Equation (10) is true only in the case in which the other six parameters are constant. It will be seen later that this relationship enables the ponding time formulae to be easily inverted. This condition does not prevail in more general cases -- for example in geometrically similar media in which case ψ_w itself is a function of K_s and therefore S varies as $K_s^{1/4}$ instead of $K_s^{1/2}$.

For determining the infiltration rate due to rainfall, we make use of the "time compression approximation" (TCA). TCA was first introduced into hydrology by Sherman (1943) and was further analyzed and tested by Ibrahim and Brutsaert (1968) and Reeves and Miller (1975). It was found to be reasonably accurate and useful in many cases. A good illustration of the application of TCA to constant rainfall infiltration is found in Milly and Eagleson (1982). Essentially, TCA implies that the infiltration capacity at any given time depends only on the previous cumulative infiltration volume, regardless of the previous rainfall history. In order to predict the infiltration capacity resulting from any surface boundary condition, one only has to determine the infiltration capacity corresponding to a simpler boundary condition for which a solution exists, provided the cumulative infiltration volume is held equal in both cases. This means that the relationship between the infiltration capacity g^* and the actual infiltration volume G for rainfall infiltration can be assumed to be the same as that for ponded infiltration. The latter can be easily obtained from Equation (7).

We applied TCA to the case of a constant rainfall rate p, with the infiltration rate formula (7), and derived the following expressions for ponding time t_p and post-ponding infiltration rate.

$$t_p = \frac{S^2}{4pcK_s} \left[\frac{p^2}{(p - cK_s)^2} - 1 \right], \quad p > K_s \tag{12}$$

$$g = cK_s + \frac{1}{2} S (t - t_c)^{-1/2}, \quad t > t_p \tag{13}$$

where $(t - t_c)$ can be defined as the "compressed" time and t_c is given by the following expression

$$t_c = \frac{S^2}{4p (p - cK_s)} \tag{14}$$

$$= \frac{S_r^2}{4p \left(\dfrac{p}{K_s} - c \right)}$$

3. CASE 1: VARIABLE SOILS, SPATIALLY UNIFORM RAINFALL

3.1 Distributions of Ponding Time

The soil hydraulic conductivity is often assumed to follow a lognormal distribution; an assumption we will follow here. The probability distribution for K_s will be denoted by $f_K(K_s)$ with mean, variance, and coefficient of variation C_{vK} given by $\exp(\mu + \frac{\sigma^2}{2})$, $\exp(2\mu + 2\sigma^2) - \exp(2\mu + \sigma^2)$, and $(\exp\sigma^2 - 1)^{1/2}$, respectively. Here μ and σ^2 are the mean and variance of $\ln K_s$. It is now possible to determine the ponding time distribution using derived distribution theory (Benjamin and Cornell, 1970).

Figure 1a is the plot of a typical lognormal probability density function of K_s and Figure 1b is the corresponding probability density function of t_p. Note in Figure 1a that a portion δ of the catchment will never be ponded. This area is represented by the shaded area for which $p < K_s$. For the limiting case $p = K_s$, the ponding time is infinite. This shaded area is expressed in the ponding time distribution, Figure 1b, as a mass equal to δ at $t_p = \infty$. The value of δ is given by

$$\delta = \frac{1}{2} - \frac{1}{2} \operatorname{erf} \left\{ \frac{(\ln p - \mu)}{\sqrt{2}\, \sigma} \right\} \tag{15}$$

In order to obtain the probability density function of t_p, it is only necessary to invert the ponding time formula (12) to express K_s as a function of t_p. That is, $K_s = h_1(t_p)$. Using (10) this is straightforward yielding

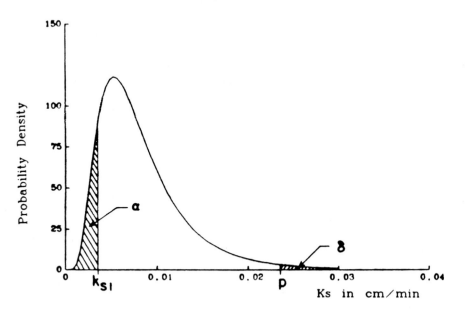

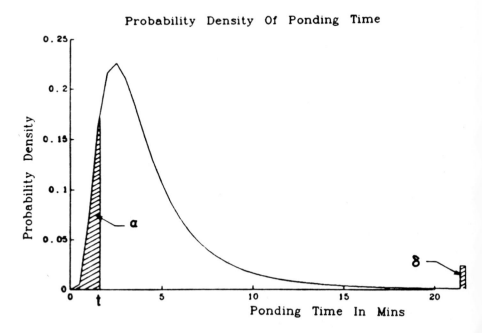

Figure 1. Illustrative sketches of probability density functions: (a) of the saturated hydr
ic conductivity K_S and (b) of the ponding time.

$$K_s \equiv h_1(t_p) = \frac{p}{c} \left[1 - \frac{\left(\dfrac{S_r^2}{4cp}\right)^{1/2}}{(t_p + \dfrac{S_r^2}{4cp})^{1/2}} \right] \qquad (16)$$

At any time t, all the soil elements having K_s values less than that value given by $h_1(t)$ would already have ponded. Let us denote the proportion of the catchment that is ponded and contributing to surface runoff at time t as $\alpha(t)$. Since K_s (and therefore $h_1(t)$) is log-normally distributed, $\alpha(t)$ can be easily shown to be

$$\alpha(t) = \frac{1}{2} + \frac{1}{2} \, \text{erf} \left\{ \frac{(\ln h_1(t) - \mu)}{\sqrt{2}\,\sigma} \right\} \qquad (17)$$

The density function of the ponding time, $f_T(t_p)$, is obtained here by derived distribution theory. Alternatively, it can also be obtained by differentiating (17), noting that $\alpha(t)$ is itself the cumulative distribution function of ponding time. The ponding time t_p is functionally related to K_s. This function can be inverted, where $f_K(K_s)$ is known, and the probability density function for the ponding time $f_T(t_p)$ can be derived as (Mood, Graybill, and Boes, 1974),

$$f_T(t_p) = \left| \frac{dh_1(t_p)}{dt_p} \right| \, f_K \, [h_1(t_p)] \qquad (18)$$

where $K_s = h(t_p)$ relates ponding time to K_s. Thus, the distribution of ponding time is

$$f_T(t_p) = \left| \frac{dh_1(t_p)}{dt_p} \right| \left[h_1(t_p)\sqrt{2\pi}\,\sigma \right]^{-1} \exp\left\{ -\frac{[\ln h_1(t_p) - \mu]^2}{2\sigma^2} \right\} \qquad (19)$$

and using (16), this can be rewritten as

$$f_T(t_p) = \frac{\dfrac{1}{2} \left(\dfrac{S_r^2}{4cp}\right)^{1/2} (t_p + \dfrac{S_r^2}{4cp})^{-1}}{\sqrt{2\pi}\,\sigma \left\{ (t_p + \dfrac{S_r^2}{4cp})^{1/2} - \left(\dfrac{S_r^2}{4cp}\right)^{1/2} \right\}}$$

$$\times \exp\left\{ -\frac{[\ln\{\dfrac{p}{c} - \dfrac{p}{c}(\dfrac{S_r^2}{4cp})^{1/2}(t_p + \dfrac{S_r^2}{4cp})^{-1/2}\} - \mu]^2}{2\sigma^2} \right\} \qquad (20)$$

We have plotted the cumulative distribution of ponding time for various cases of rainfall p and coefficient of variation C_{vK}. These results correspond to a mean soil hydraulic conductivity, $\overline{K}_s$, equal to 0.008325 cm/min, S_r equal to 1.9837, and c equal to 0.667; these values of $\overline{K}_s$, S_r, and c are used in all the examples presented in this paper. Figure 2 gives the cumulative distribution of ponding times for a rainfall rate of $p = 0.01665$ cm/min while Figure 3 gives the corresponding functions for $p = 0.0333$ cm/min. In these figures, the cumulative distributions of ponding times, denoted by $\alpha(t)$, are given as functions of time for different values of C_{vK}. These figures also help to understand the effect of spatial variability in soil properties on the proportion of the catchment contributing to surface runoff.

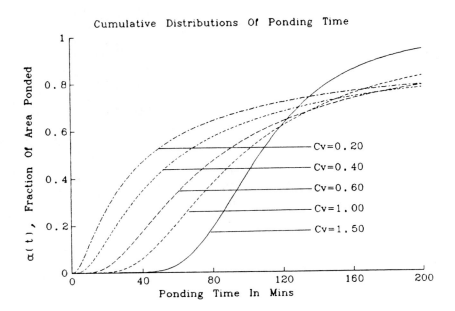

Figure 2. Cumulative distribution functions of ponding time due to variable soils as a function of C_{VK} for the case p = 0.01655 cm/min and $\overline{K}_S$ = 0.008325 cm/min.

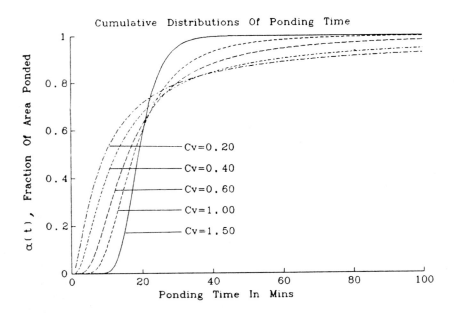

Figure 3. Cumulative distribution functions of ponding time due to variable soils as a function of C_{VK} for the case p = 0.0333 cm/min and $\overline{K}_S$ = 0.008325 cm/min.

It can be seen from Figures 2 and 3, for example, that as C_{vK} increases, the threshold between rainfall controlled infiltration and soil controlled infiltration becomes less distinct as when compared to the point infiltration response.

To see this more clearly, consider the case of how soil variability, i.e., $C_v = 0.20$. For the low rainfall rate, virtually no ponding takes place before 40 minutes. Doubling the rainfall rate (Figure 3) results in reducing this time to first appearance of ponding by a factor of 4, that is, no ponding has occurred before about 10 minutes. For highly variable catchments $(C_v = 1.50)$ ponding is initiated at some locations on the catchment almost immediately for both rainfall cases.

One can also observe that the expansion of contributing areas occurs more quickly in relatively homogeneous catchments, once ponding is initiated, than in heterogeneous catchments. On the other hand, at low rainfall rates and short rainfall periods, heterogeneous catchments almost always appear to have ponded areas and therefore runoff generating areas. For relatively homogeneous areas, many storms may not result in runoff production areas.

3.2 Mean Infiltration Rate

The mean infiltration rate can be expressed as

$$m_g(t) = \int_0^\infty g(t;K_s) \, f_K(K_s) \, dK_s \tag{21}$$

where $f_K(K_s)$ is the distribution of K_s. This equation can be expanded as

$$m_g(t) = [1 - \alpha(t)]p$$
$$+ \int_0^{K_{s1}=h_1(t)} \left\{ cK_s + \frac{S_r K_s^{1/2}}{2\left[t - \dfrac{S_r^2}{4p\left(\dfrac{p}{K_s} - c\right)}\right]^{1/2}} \right\} f_K(K_s) \, dK_s \tag{22}$$

The first part of Equation (22) states that at time t, $(1 - \alpha)$ portion of the catchment is yet to start ponding and thus has infiltration rates equal to p. For the remainder of the catchment, which has saturated conductivity values less than that given by $h_1(t)$, ponding has occurred and the infiltration rate at any point is given by Equation (13). We now approximate the solution to Equation (22) by the following equation:

$$m_g(t) = \left[1 - \alpha(t)\right]p + c\overline{K_s}^\alpha + \frac{S_r \overline{K_s^{1/2^\alpha}}}{2\left[t - \dfrac{S_r^2}{4p\left(\dfrac{p}{\alpha \overline{K_s}} - c\right)}\right]^{1/2}} \tag{23}$$

where

$$\overline{K_s^{m^\alpha}} = \int_0^{h_1(t)} K_s^m \, f_K(K_s) \, dK_s \tag{24}$$

and where we have effectively removed the term involving the compressed time $(t - t_c)$ from the integral by replacing K_s in t_c by $\alpha \overline{K_s}$ (see Eqn. 14). The basis for this approximation is our conviction that at most times t, the value of t_c is small compared to t, and therefore, the term inside the integral involving $(t - t_c)$ can be removed from the integral by using an appropriate averaged t_c or an averaged K_s. While we consider the approximation used in Equation (25) as adequate for our purposes here, the influence of the approximation on the mean infiltration rate will have to be evaluated more fully. Since the probability density function $f_K(K_s)$ is assumed lognormal, (24) can be solved to yield

$$\overline{K_s^{m\alpha}} = \exp\left\{ m\mu + \frac{m^2\sigma^2}{2} \right\} \left\{ \frac{1}{2} + \frac{1}{2} \operatorname{erf}\left[\frac{(\ln h_1(t) - \mu - m\sigma^2)}{\sqrt{2}\,\sigma} \right] \right\} \tag{25}$$

Equation (23), for the mean areal infiltration rate, presents a very important, but obvious, qualitative result. It says that at any time t, the average infiltration rate is a weighted summation of the rainfall controlled infiltration over the nonponded portion and the mean soil controlled infiltration over the ponded portion, where the weighting parameter is $\alpha(t)$.

Mean infiltration rates, based upon Equation (23), are presented in Figure 4 for rainfall rate p = 0.01665 cm/min and in Figure 6 for p = 0.0333 cm/min. The point infiltration response for a spatially constant soil with K_s equal to the areal mean $\overline{K}_s$ is given by the curve denoted by $C_{vK} = 0.0$. The curves indicate that considerable bias arises when one attempts to model the mean areal response of a catchment with spatially variable soil properties by using averaged soil properties in conjunction with the point infiltration equation. This bias increases as the spatial variability increases, as characterized by increasing C_{vK}.

As mentioned before, the cumulative distribution of ponding time, which is equal to the proportion of the area that is ponded, is a convenient variable that helps in the parameterization of mean areal infiltration for spatially variable soils. The validity of the approximations made in deriving Equation (23) was assessed by comparing the results obtained by the approximate formula with those arising from a Monte Carlo simulation. We found good agreement between the two values.

3.3 Variance of the Infiltration Rate

The equation for the variance of the infiltration rate can be expressed as

$$\sigma_g^2(t) = \int_0^\infty g^2(t;K_s)\, f_K(K_s)\,dK_s - m_g^2(t) \tag{26}$$

We then have

$$\sigma_g^2(t) = (1-\alpha)p^2 - m_g^2(t) + c^2 \int_0^{h_1(t)} K_s^2\, f_K(K_s)\, dK_s$$

$$+ \int_0^{h_1(t)} \frac{S_r^2 K_s}{4\left[t - \dfrac{S_r^2}{4p\left(\dfrac{p}{K_s} - c\right)} \right]}\, f_K(K_s)\,dK_s$$

$$+ c \int_0^{h_1(t)} \frac{S_r K_s^{3/2}}{\left[t - \dfrac{S_r^2}{4p\left(\dfrac{p}{K_s} - c\right)} \right]}\, f_K(K_s)\,dK_s \tag{27}$$

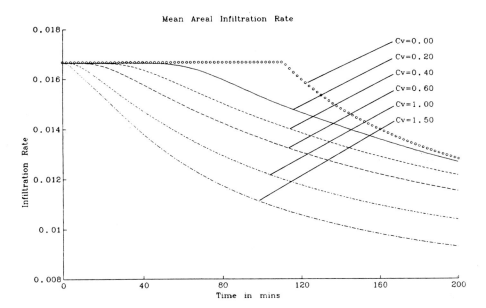

Figure 4. Mean areal infiltration rate due to variable soils as a function of C_{VK} for the case p $= 0.01665$ cm/min and $\overline{K}_S = 0.008325$ cm/min.

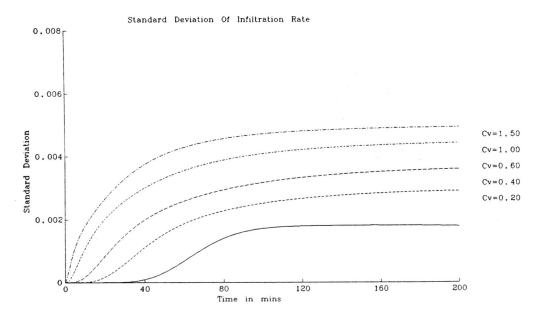

Figure 5. Standard deviation of the infiltration rate due to vriable soils as a function of C_{VK} for the case p $= 0.01665$ cm/min and $\overline{K}_S = 0.008325$ cm/min.

By removing the terms involving $(t - t_c)$ from the integral by replacing K_s in t_c by $\alpha \overline{K}_s$, we have

$$\sigma_g^2(t) = (1 - \alpha)p^2 - m_g^2(t) + c^2 \, \overline{K_s^{2\alpha}}$$

$$+ \frac{S_r^2 \, \overline{K_s^{\alpha}}}{4 \left[t - \dfrac{S_r^2}{4p \left(\dfrac{p}{\alpha \overline{K}_s} - c \right)} \right]} + \frac{cS_r \, \overline{K_s^{3/2\alpha}}}{\left[t - \dfrac{S_r^2}{4p \left(\dfrac{p}{\alpha \overline{K}_s} - c \right)} \right]} \tag{28}$$

The values of the standard deviation calculated by Equation (28) are presented in Figure 5 for rainfall rate p = 0.01665 cm/min and in Figure 7 for p = 0.0333 cm/min. The standard deviation is zero at time $t = 0$, increases at early time, and finally, for large time, approaches the value of the underlying K_s random field. We compared the standard deviations calculated from the above equation to those obtained by Monte Carlo simulation and found that the agreement was reasonably good.

3.4 Covariance of Infiltration Rate

The covariance function of the infiltration rate random field can be obtained by considering two arbitrary points in space separated by a distance $r = |\underset{\sim}{r}|$. The covariance is given by

$$C_g(r,t) = \int\limits_{-\infty}^{+\infty} \int\limits_{-\infty}^{+\infty} g(t;y_1)g(t;y_2)f_{Y_1 Y_2}(y_1,y_2)dy_1 dy_2 - m_g^2(t) \tag{29}$$

where $f_{Y_1 Y_2}(y_1,y_2)$ is the joint probability density function of $y_1 = \ln(K_s(\underset{\sim}{x}))$ and $Y_2 = \ln(K_s(\underset{\sim}{x} + \underset{\sim}{r}))$ and is a binormal distribution;

$$f_{Y_1 Y_2}(y_1,y_2) = \frac{1}{2\pi\sigma^2\sqrt{1-\rho^2}} \tag{30}$$

$$\exp\left\{ -\frac{1}{2\sigma^2(1 - \rho^2)} \left[(y_1 - \mu)^2 - 2\rho(y_1 - \mu)(y_2 - \mu) + (y_2 - \mu)^2 \right] \right\}$$

where ρ is the correlation coefficient for $\ln(K_s)$ between the two chosen points a distance r apart.

Only limited field investigations have been performed to determine the actual correlation structure of soil properties; both the exponential and the Bessel forms have been suggested. Whittle (1954) argued that a Bessel form is more suitable for two dimensional isotropic processes; following Whittle we have used the following Bessel type correlation function.

$$\rho_{\ln(K_s)}(r) = brK_1(br), \quad r = |\underset{\sim}{r}| \tag{31}$$

where K_1 is the modified Bessel function of the second kind and of order one. The constant b and the integral scale are related by

$$\lambda_{\ln(K_s)} \equiv \int\limits_0^{\infty} \rho_{\ln(K_s)}(r)dr = \frac{\Pi}{2b} \tag{32}$$

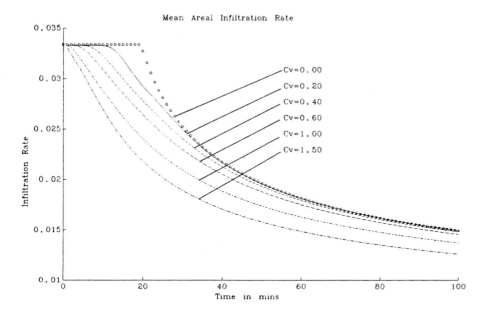

Figure 6. Mean areal infiltration rate due to variable soils as a function of C_{VK} for the case p $= 0.0333$ cm/min and $\bar{K}_S = 0.008325$ cm/min.

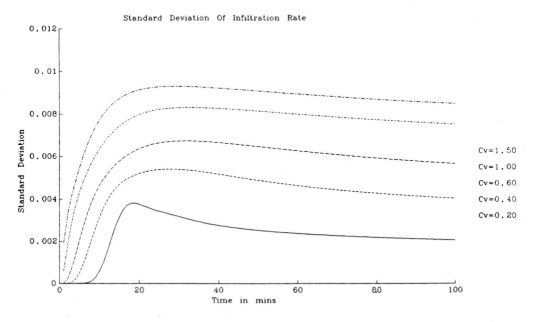

Figure 7. Standard deviation of the infiltration rate due to variable soils as a function of C_{VK} for the case p$= 0.0333$ cm/min and $\bar{K}_S = 0.008325$ cm/min.

The relationship between the correlations of the $\ln(K_s)$ and K_s processes has been derived by Mejia and Rodriguéz-Iturbe (1974).

Based on the Bessel correlation function, we derived the covariance function of the infiltration rate and approximated the resulting expressions by taking the terms involving $(t - t_c)$ outside of the integrals by replacing K_s in t_c by an appropriately averaged K_s. We will not present the derivations here, noting only that the resulting expressions finally involved the evaluation of probability integrals of binormal distributions over different rectangular base areas; the integrals were determined using numerical approximations. Once the infiltration covariance is determined, the correlation is given by

$$\rho_g(r,t) = \frac{C_g(r,t)}{\sigma_g^2(t)} \tag{33}$$

The correlations calculated from the equations derived in this section are presented in Figures 8 and 9. All calculations were performed for a K_s field having mean value equal to 0.008325 cm/min. The time variations of the spatial correlograms for the infiltration rate are also presented in Figures 8 and 9. Figure 8 presents the correlograms for p = 0.01665 cm/min while Figure 9 gives the resulting infiltration rate correlograms for p = 0.0333 cm/min.

An important feature observed in the figures presented above is that the correlation structure of the infiltration field is always less than that of the underlying random field of K_s, except at extremely early times. At time zero, the correlation is that of the rainfall field since at very early times the infiltration is equal to the rainfall. However, as time increases the correlation drops rapidly to a small value and then starts increasing and at large time tends to the value of the underlying K_s field. Consequently, except at very short times, the scale of variability of infiltration rate is always less than that of the soil hydraulic conductivity. This observed result arises by the fact that the rainfall does not contribute to the infiltration covariance even though the rainfall correlation in space is 1.0, because the variance of the rainfall rate is zero.

4. CASE 2: VARIABLE RAINFALL, UNIFORM SOIL PROPERTIES

The derivations of the distributions of ponding time and the moments of the infiltration rate for the variable rainfall case are essentially the same as before. We assume that the rainfall intensities follow a two parameter gamma distribution characterized by parameters ν and ω and with mean $\omega\nu$ and variance $\omega^2\nu$.

4.1 Distribution of Ponding Time

As before, the distribution of ponding time is obtained by derived distribution theory. The first task is to invert Equation (12) to express the rainfall intensity p as a function of the ponding time t_p; that is

$$p = h_2(t_p) \tag{34}$$

For the variable rainfall case, the inversion cannot be obtained analytically but will have to be obtained numerically. Once this is done the derived distribution follows in a straightforward manner. The resulting pdf of the ponding time t_p is

$$f_T(t_p) = \frac{\omega^{-\nu}}{\Gamma(\nu)} \exp(-\frac{h_2(t_p)}{\omega}) h_2^{\nu-1}(t_p)$$

$$\times \left\{ 1 - \frac{S_r^2 K_s}{64} \frac{\left[8h_2(t_p) - 3K_s\right]^2 + 7K_s^2}{h_2^2(t_p)[h_2(t_p) - K_s]^3} \right\}^{-1} \tag{35}$$

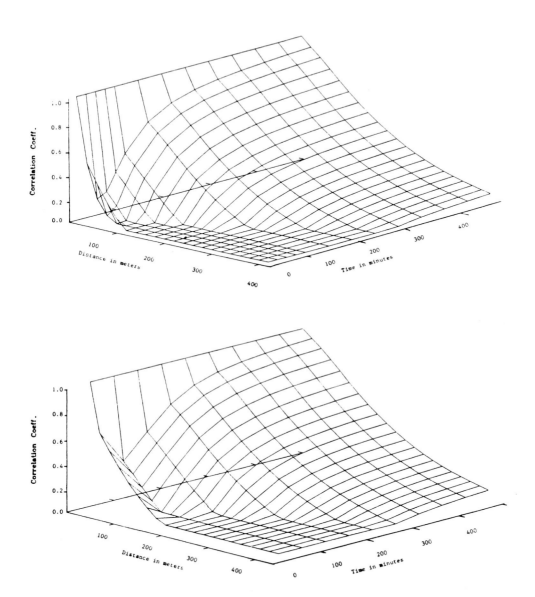

Figure 8. Space correlations of infiltration rate, due to spatially variable soils, as a function of time and space for the case p $= 0.01665$ cm/min and $\overline{K}_S = 0.008325$ cm/min: (a) $C_{VK} = 0.4$ and (b) $C_{VK} = 0.8$. For all cases $\lambda_{lnK_S} = 200$m.

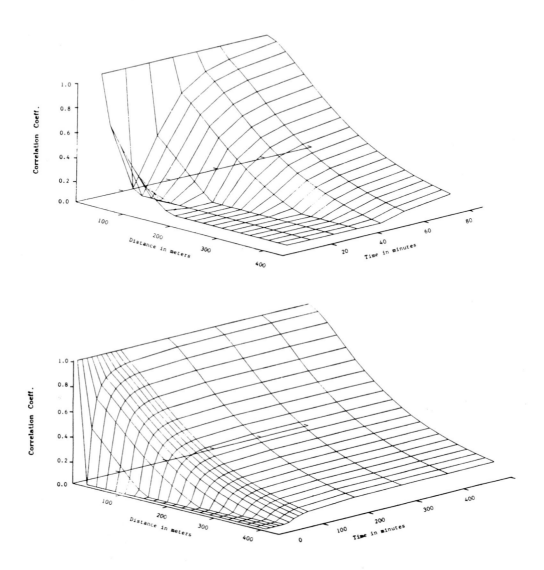

Figure 9. Space correlations of infiltration rate, due to spatially variable soils, as a function of time and space for the case p $= 0.0333$ cm/min and $\overline{K}_S = 0.008325$ cm/min. (a) $C_{VK} = 0.4$ and (b) $C_{VK} = 1.6$. For all cases $\lambda_{lnK_S} = 200$m.

The cumulative distribution of t_p which is identical to the proportion of catchment that is ponded is given by

$$\alpha(t) = 1 - \int_0^{p = h_2(t)} f_P(p)\,dp \tag{36}$$

where the term involving the integral is a cumulative distribution of the gamma distribution and is computed using the incomplete gamma function.

The cumulative distributions of the ponding time were determined for various values of mean rainfall intensity and coefficient of variation C_{vp}. These are shown in Figures 10 and 11. All calculations for the variable rainfall case were carried out for $K_s = 0.008325$ cm/min, $S_r = 1.9837$ and $c = 0.667$.

4.2 Mean Infiltration Rate

The mean infiltration rate in the case of spatially variable rainfall can be determined by the same procedure adopted for the previous case.

$$m_g(t) = \int_0^{p = h_2(t)} p f_P(p)\,dp + c\,\alpha K_s \tag{37}$$

$$+ \frac{S_r}{2} K_s^{1/2} \int_{p = h_2(t)}^{\infty} \frac{1}{(t - t_c)^{1/2}} f_P(p)\,dp$$

The first term on the right hand side can be shown to be the cumulative of a gamma distribution, and can be evaluated exactly. The integral in the last term on the right can be approximated, as before, by approximating the term involving $(t - t_c)$ and removing it from the integral. The approximate equation for the mean infiltration rate is given by

$$m_g(t) = \bar{p}^{(1 - \alpha)} + c\,\alpha K_s + \frac{S_r K_s^{1/2}\alpha}{2} \frac{1}{(t - \bar{t_c}^\alpha)^{1/2}} \tag{38}$$

where

$$\bar{t_c}^\alpha = S_r^2 K_s \left[4\bar{p}^\alpha(\bar{p}^\alpha - cK_s) \right]^{-1} \tag{39}$$

In Equations (38) and (39) $\bar{p}^\alpha$ and $\bar{p}^{(1 - \alpha)}$ are the integrals of p over the ponded and non-ponded regions respectively. The mean infiltration rates were calculated using Equation (38) for various combinations of the rainfall rate statistics. These are shown in Figures 12 and 14. As before, there is a strong bias in the mean infiltration rates on a catchment with spatially variable rainfall, as when compared to the predictions obtained from the point infiltration equation using the mean rainfall intensity. This bias increases with increasing coefficient of variation, C_{vp}.

4.3 Variance of Infiltration Rate

The variance of the infiltration rate was calculated in a similar manner. The approximate equation for the variance of the infiltration rate is given by

$$\sigma_g^2(t) = \bar{p}^{2(1 - \alpha)} + \alpha c^2 K_s^2 - m_g^2(t) + \frac{S_r^2 K_s}{4(t - \bar{t_c}^\alpha)} + \frac{cS_r K_s^{3/2}}{(t - \bar{t_c}^\alpha)^{1/2}} \tag{40}$$

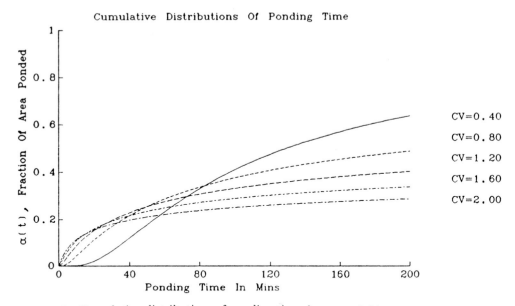

Figure 10. Cumulative distributions of ponding time due to variable rainfall for the case $\bar{p} = 0.01665$ cm/min and $K_S = 0.008325$ cm/min as a function of C_{vp}.

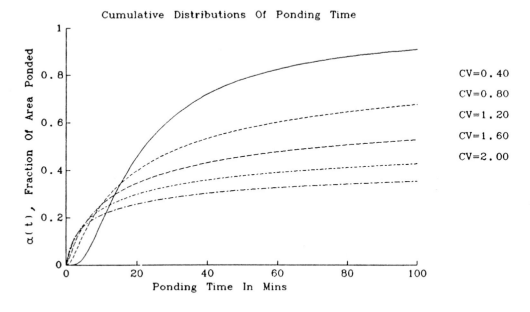

Figure 11. Cumulative distributions of ponding time due to variable rainfall for the case $\bar{p} = 0.0333$ cm/min and $K_S = 0.008325$ cm/min as a function of C_{vp}.

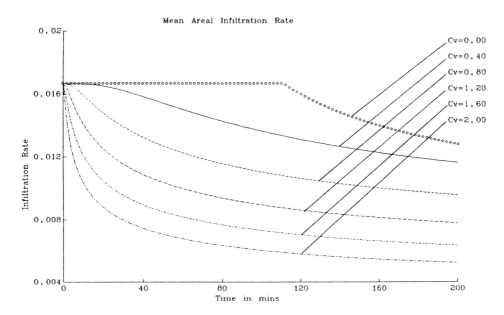

Figure 12. Mean areal infiltration rate due to variable rainfall intensity as a function of C_{vp} for the case $\bar{p} = 0.01665$ cm/min and $K_S = 0.008325$ cm/min.

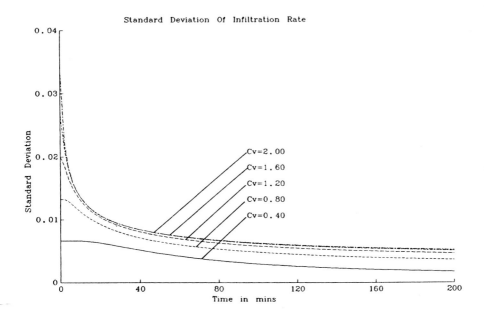

Figure 13. Standard deviation of infiltration rate due to variable rainfall intensity as function of C_{vp} for the case $\bar{p} = 0.01665$ cm/min and $K_S = 0.008325$ cm/min.

where $\overline{p^{2}}^{1-\alpha}$ is the integral of p^2 over the nonponded region of the catchment. The standard deviations of the infiltration rate for various combinations of mean rainfall and C_{vp} are shown in Figures 13 and 15. The standard deviation of the infiltration rate falls rapidly from being equal to that of the rainfall initially, to a fairly steady, small but nonzero, value at large times.

The mean and standard deviation of the infiltration rate were also calculated using more accurate numerical methods and they showed the suitability of the approximations adopted in the derivations.

4.4 Covariance of Infiltration Rate

The covariance of the infiltration rate due to spatially variable rainfall is given by

$$C_g(r,t) = \int_0^\infty \int_0^\infty g(t;p_1)g(t;p_2)f_{P_1 P_2}(p_1,p_2)dp_1 dp_2 - m_g^2(t) \tag{41}$$

where $f_{P_1 P_2}(p_1,p_2)$ is the bivariate pdf of rainfall and is represented by a bivariate gamma distribution. The bivariate gamma distribution is given by Nagao and Kadoya (1971) as

$$f_{P_1 P_2}(p_1,p_2) = \frac{1}{\Gamma(\nu)\omega^{\frac{\nu+1}{2}}(1-\rho)\rho^{\frac{\nu-1}{2}}} \exp\left\{-\frac{p_1}{\omega(1-\rho)} - \frac{p_2}{\omega(1-\rho)}\right\} \tag{42}$$

$$\times (p_1,p_2)^{\frac{\nu-1}{2}} I_{\nu-1}\left[\frac{2\sqrt{\rho}}{1-\rho}\left(\frac{p_1 p_2}{\omega^2}\right)^{0.5}\right]$$

where ρ is the correlation between values of p at two points $\underset{\sim}{x}$ and $\underset{\sim}{x} + \underset{\sim}{r}$, separated by a distance $|\underset{\sim}{r}|$, and $I_{\nu-1}$ is the modified Bessel function or order $(\nu-1)$.

The spatial correlation structure of the rainfall field is assumed to be isotropic and given by

$$\rho_p(r) = a_1 \exp\left(-b_1^2 r^2\right) + a_2 \exp\left(-b_2^2 r^2\right) \tag{43}$$

where a_1, a_2, b_1, and b_2 are constants such that $a_1 + a_2 = 1.0$. The combination of two or more correlation structures of different integral scales has been shown to reproduce the cellular or banded structure of actual rainfall correlation structures (Journel and Huijbregts, 1978; Valdes, Rodriguez-Iturbe and Gupta, 1985). The integral scale of the rainfall correlation is given by

$$\lambda_p = \frac{a_1\sqrt{\pi}}{b_1} + \frac{a_2\sqrt{\pi}}{b_2} \tag{44}$$

The covariance of the infiltration rate was obtained by the numerical integration of Equation (41). The resulting space correlations of the infiltration rate are plotted as functions of time in Figures 16 and 17. Initially, the correlation of the infiltration rate is equal to that of rainfall. In contrast to the variable soils case, the correlation of the infiltration rate decreases slowly and after many hours becomes more or less steady. The interpretation for this is that the covariance of K_s is zero and does not contribute to the covariance of infiltration after the latter has become soil controlled.

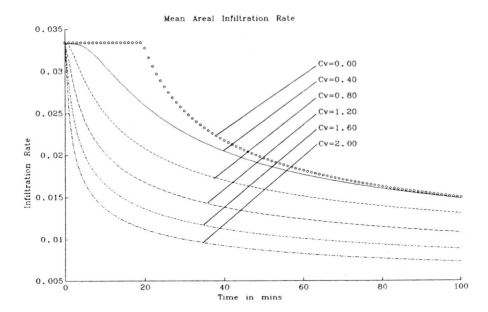

Figure 14. Mean areal infiltration rate due to variable rainfall intensity as a function of C_{vp} for the case $\bar{p}$ = 0.0333 cm/min and K_S = 0.008325 cm/min.

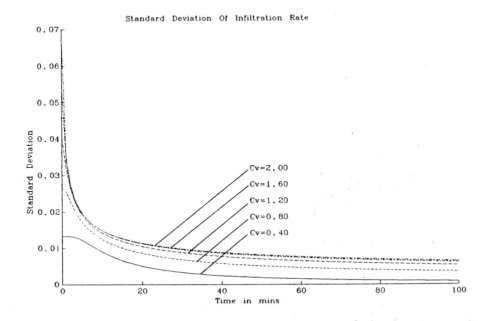

Figure 15. Standard deviation of infiltration rate due to variable rainfall intensity as a function of C_{vp} for the case $\bar{p}$ = 0.0333 cm/min and K_S = 0.008325 cm/min.

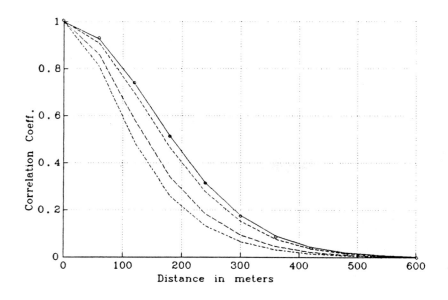

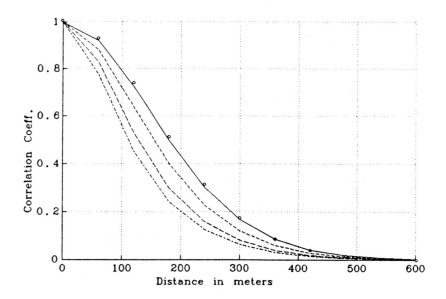

Figure 16. Space correlations of the infiltration rate due to spatially variable rainfall as a function of time and space for the case $\bar{p} = 0.01665$ cm/min and $K_S = 0.008325$ cm/min: (a) $C_{vp} = 1.00$ and (b) $C_{vp} = 1.60$. In all cases $\lambda_p = 200$m.

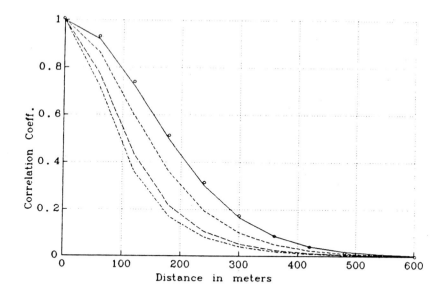

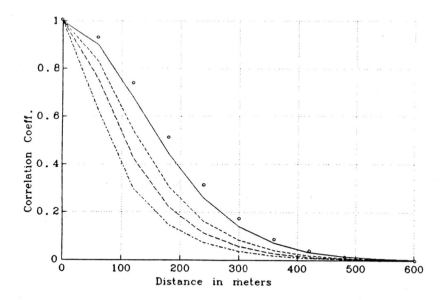

Figure 17. Space correlations of the infiltration rate due to spatially variable rainfall as a function of time and space for the case $\bar{p} = 0.0333$ cm/min and $K_S = 0.008325$ cm/min: (a) $C_{vp} = 1.00$ and (b) $C_{vp} = 1.60$. In all cases $\lambda_p = 200$m.

5. DISCUSSION

We have derived expressions for the mean areal infiltration rate and its covariance for two different situations: (1) spatially variable soils and constant rainfall, and (2) spatially variable rainfall and constant soils. The results showed that considerable bias would arise if only mean values are used in conjunction with the point infiltration equation.

However, the above derivations have implicitly made use of the ergodic assumption; namely, that the averaging area is infinitely large so that the areal mean is identically equal to the ensemble mean. In practice, actual catchments are probably not large enough to justify the ergodic assumption. In such cases the actual correlation scales of the infiltration rate and runoff production become very important.

The space correlations for the infiltration rate were calculated for the two situations mentioned above. The results indicate opposing roles for the rainfall and soil variabilities in runoff production. In the case of variable soils, with zero rainfall covariance, the infiltration rate correlation at all times is controlled by the covariance of K_s and $\alpha(t)$. Similarly, in the case of variable rainfall, the infiltration correlation is controlled by the rainfall covariance and $\alpha(t)$. In an actual catchment, where the variances of both rainfall and soils are nonzero, one would expect that both rainfall and soil variabilities would play opposing roles in determining the actual scale of variability of the infiltration rate. The relative importance of rainfall or soil variabilities would be expected to depend on their respective covariances. This is discussed in more detail in Wood et al. (1986).

In the determination of the mean areal infiltration rate, for the ergodic assumption to hold, the catchment area should be at least a certain multiple of the square of the infiltration rate correlation length. Based on the results obtained in this paper, it can be observed that the required catchment area is much larger for the variable rainfall case than for the variable soils case. Thus, it is evident that rainfall variability is the more critical of the two in modeling the rainfall-runoff response. Also, one may expect that once the catchment area exceeds a certain threshold, which will be governed by the scale of variability of K_s, the soil properties can essentially be considered as random and independent. Work is in progress which will more fully analyze these concepts.

6. CONCLUSION

The aim of the paper was to understand the effect of spatial heterogeneity on the infiltration response of catchments with either spatially variable soils or rainfall. The approach taken was to derive quasi-analytical expressions for the distributions of ponding times, the mean infiltration rate, and the covariance structure.

Conceptually, the two cases analyzed in this paper represent two extreme cases in how catchment and input variability can affect infiltration responses. One could approximate the two cases to different scales of analysis.

For case 1, constant rainfall and variable soils may relate to medium sized, geologically variable catchments subject to large cyclonic events of relatively low rainfall intensities. Thus, the rainfall correlation length is much longer than those for the soil hydraulic conductivity.

Case 2, variable rainfall and constant soils could be related to small to medium homogeneous catchments subject to convective rain events. Here the rainfall correlation length is much smaller than those of the soil properties.

The derivations showed that the critical variable in understanding the spatial response is the time to ponding and its distribution in time and space. The cumulative distribution functions for time to ponding, $\alpha(t)$, developed in Equation (17) for the spatially variable soil case and in Equation (36) for the spatially variable rainfall case, indicate that portion of the catchment where infiltration is soil controlled and that portion that is rainfall controlled.

The results showed that for spatially variable soils and rainfall the use of average soil properties, or rainfall intensity, leads to biased mean areal infiltration responses. The appropriate mean infiltration equation for spatially variable soils is Equation (23) and for spatially variable rainfall is Equation (38).

Finally, the covariance of the infiltration rate is found to be a function of space and time in both cases, with the rainfall and soil correlations acting as upper and lower bounds respectively for the correlation of the infiltration rate. This result is due to the heterogeneity of the infiltration response; part of the catchment being soil controlled and the remainder being rainfall controlled. The results indicate opposing roles for rainfall and soil variabilities; the importance of each would depend on the relative magnitude of its variance. Since the rainfall correlation is usually larger than that of K_s, the former is expected to play a more critical role in rainfall-runoff modeling.

ACKNOWLEDGMENT

This research was supported in part by NSF Grant CEE-8100491 and by NASA Grant NAG-5/491.

REFERENCES

Benjamin, J.R., and C.A. Cornell, 1970, *Probability, Statistics, and Decision for Civil Engineers*, McGraw-Hill, New York.

Bresler, E., and G. Dagan, 1983, "Unsaturated Flow in Spatially Variable Fields, 2. Application of Water Flow Models to Various Fields", *Water Resources Research*, Vol. 19, No. 2, pp. 421-428.

Brutsaert, W. 1976, "The Concise Formulation of Diffusive Sorption of Water in Dry Soil", *Water Resources Research*, Vol. 12, No. 6, pp. 1118-1124.

Dagan, G., and E. Bresler, 1983, "Unsaturated Flow in Spatially Variable Fields, 1. Derivation of Models of Infiltration and Redistribution", *Water Resources Research*, Vol. 19, No. 2, pp. 413-420.

Dooge, J.C.I., 1981, "Parameterization of Hydrologic Processes", paper presented at the JSC Study Conference on Land Surface Processes in Atmospheric General Circulation Models, Greenbelt, USA, January, pp. 243-284.

Ibrahim, H.A. and W. Brutsaert, 1968, "Intermittent Infiltration into Soils with Hysteresis", ASCE, *Journal of Hydrology*, HY1, pp. 113-137.

Journel, A.G., and Ch. J. Huijbregts, 1978, *Mining Geostatistics*, Academic Press, 600 pp.

Maller, R.A., and M.L. Sharma, 1981, "An Analysis of Areal Infiltration Considering Spatial Variability", *Journal of Hydrology*, Vol. 52, pp. 25-37.

Maller, R.A. and M.L. Sharma, 1984, "Aspects of Rainfall Excess from Spatially Varying Hydrological Parameters", *Journal of Hydrology*, Vol. 67, pp. 115-127.

Mijia, J.M., and I. Rodríguez-Iturbe, 1974, "Correlation Links Between Normal and Lognormal Processes", *Water Resources Research*, Vol. 10, No. 4, pp. 689-690.

Milly, P.C.D. and P.S. Eagleson, 1982, "Infiltration and Evaporation at Inhomogeneous Land Surfaces", MIT Ralph M. Parsons Laboratory Report No. 278, 180 pp.

Mood, A.M., F.A. Graybill, and D.C. Boes, 1974, *Introduction to the Theory of Statistics*, McGraw-Hill, Third Edition, 564 pp.

Nagao, M., and M. Kadoya, 1971, "Two Variate Exponential Distribution and its Numerical Table for Engineering Application", *Bull. Disas. Prev. Res. Inst.*, Kyoto University, Vol. 20, Part 3, No. 178, pp. 183-197.

Philip, J.R., 1957, "The Theory of Infiltration", *Soil Science*, Vols. 83, 84 and 85.

Reeves, M. and E.E. Miller, 1975, "Estimating Infiltration for Erratic Rainfall", *Water Resources Research*, Vol. 11, No. 1, pp. 102-110.

Rodríguez-Iturbe, I., and J.M. Mejia, 1974, "On the Transformation from Point Rainfall to Areal Rainfall", *Water Resources Research*, Vol. 10, No. 4, pp. 729-735.

Sharma, M.L., and E. Seely, 1979, "Spatial Variability and its Effects on Areal Infiltration", *Proceedings Hydrology and Water Resources Symposium*, Inst. Eng., Australia, Perth, W.A., pp. 69-73.

Sherman, L.K., 1943, "Comparison of F-curves Derived by the Methods Sharp and Holtan and of Sherman and Mayer", *Trans. Am. Geophys. Un.*, Vol. 24, pp. 465-467.

Smith, R.E., and R.H.P. Hebbert, 1979, "A Monte Carlo Analysis of the Hydrologic Effects of Spatial Variability of Infiltration", *Water Resources Research*, Vol. 15, No. 3, pp. 419-429.

Valdes, J.B., I. Rodríguez-Iturbe, and V.K. Gupta, 1985, "Approximations of Temporal Rainfall from a Multidimensional Model", *Water Resources Research*, Vol. 21, No. 8, pp. 1259-1270.

Whittle, P., 1954, "On Stationary Processes in the Plane", *Biometrika*, Vol. 41, pp. 434-449.

Wood, E.F., M. Sivapalan, and K. Beven, 1986, "Scale Effects in Infiltration and Runoff Production", Paper to be presented at the 2nd Scientific Assembly of the IAHS, Budapest Hungary, July 2-10, 1986.

RUNOFF PRODUCTION AND FLOOD FREQUENCY IN CATCHMENTS OF ORDER n: AN ALTERNATIVE APPROACH

6

Keith Beven

ABSTRACT

A simple physically-based hydrological model is derived that takes account of the effect of spatial heterogeneities of topography and soil on runoff production. Both infiltration excess and saturation excess mechanisms of runoff production are simulated by the model. The model is computationally inexpensive and has been used to derive flood frequency characteristics for three small catchments by simulating hydrographs during a 100 year record of randomly generated rainstorms. Interstorm calculations are carried out analytically. For the range of parameter values studied it was found that all the maximum annual flood peak distributions are of extreme value 1 (Gumbel) type. The normalized distribution functions (growth curves) are remarkably similar over all the parameter sets considered. Runoff production in flood events for all the simulations was dominated by the saturation excess mechanism, even assuming very high hydraulic conductivities, and even where infiltration excess runoff is predicted as occurring over part of the catchment. It appears to be difficult to avoid surface saturation under the wet conditions associated with floods in a climatic regime typical of upland Britain.

1. INTRODUCTION

This paper is concerned with the prediction of the flood frequency characteristics of catchments of different scales, particularly ungauged catchments. Traditionally this problem has been tackled by the use of regionalized relationships between flood frequency characteristics and catchment characteristics. An example is the UK Flood Study Report (NERC, 1975). With some exceptions, such studies have included catchments of different size in the same regression analysis, the flood frequency curves often being normalized by being expressed relative to the mean annual flood. This assumes an underlying shape of growth curve or normalized distribution function that is independent of catchment scale or characteristics. There is

107

V. K. Gupta et al. (eds.), Scale Problems in Hydrology, 107–131.
© *1986 by D. Reidel Publishing Company.*

no a priori reason to accept this assumption, but it is worth noting that the UK Flood Study found no reason to stratify on the basis of catchment scale. They suggested that "geology and climate can play a larger part than area alone." (v.I, p. 182), demonstrating a wide range of variability in the growth curves for small catchments.

For the prediction of mean annual flood the Flood Study report found that:

> "little would be gained by dividing the records into ranges of catchment size; on the other hand, significant improvements could be made by dividing the country into geographical regions" (v. I, p. 11).

They also comment that

> "an interesting byproduct of these tests is the poor fit obtained for smaller catchments; larger catchments would appear to behave more regularly" (v. I, p. 329).

These conclusions would suggest that it should be possible to improve upon the regression methodology by taking a more physically-based approach to flood frequency prediction. The work reported here has been stimulated by the recent contributions toward this end by Hebson and Wood (1982), Cordova and Rodriguéz- Iturbe (1983) and Diaz-Granados et al. (1984), building on the pioneering work of Eagleson (1972).

The recent papers have all used variations on the idea of the geomorphological unit hydrograph (GUH) to transform "excess rainfall" into the discharge hydrograph. The geomorphological unit hydrograph is an attempt to express the way that runoff routing is dependent on catchment scale and structure, and in recent forms (e.g., Rodriguéz- Iturbe et al. 1982) also on rates of runoff production. The papers differ in the way that they estimate runoff production. Hebson and Wood (1982) used a constant partial contributing area, Cordova and Rodriguéz-Iturbe (1983) used the SCS curve number technique, and Diaz-Granados et al. (1984) used the Philip infiltration equation.

This paper concentrates on the processes of runoff production. The author agrees with Cordova and Rodriguéz-Iturbe (1983) that "the problem is more what to route than how to route...what remains as a crucial and unsolved problem is the description of the infiltration process at a basin scale." (p. 172). None of the techniques described above are satisfactory descriptions of the runoff production process at the basin scale. Indeed misconceptions about mechanisms of runoff production remain widespread, despite the excellent review of Dunne (1978). This paper attempts to introduce an integrated model of several different mechanisms of runoff production into a physically- based model of the flood frequency characteristics for catchments of different scales. As such, this work extends the work reported in Beven (1985) which was based only on a saturation excess runoff production mechanism.

2. MECHANISMS OF RUNOFF PRODUCTION

We envisage a spectrum of processes that may be involved in flood runoff production. Some members of this spectrum are illustrated in Figure 1. The response of any particular basin may be dominated by a single mechanism, or more usually may involve 2 or more mechanisms that may occur at the same time in different parts of a basin. The balance of response processes in any particular basin may be dependent on antecedent conditions, local storm rainfall rates and heterogeneities in soil hydraulic properties, as illustrated for example by the simulations of Freeze (1980). Freeze however failed to predict the widespread occurrence of saturation excess runoff that Dunne (1978) suggests should be found in humid temperate areas. This is because his simulation did not include convergent topographic hollows that are important in inducing surface saturation.

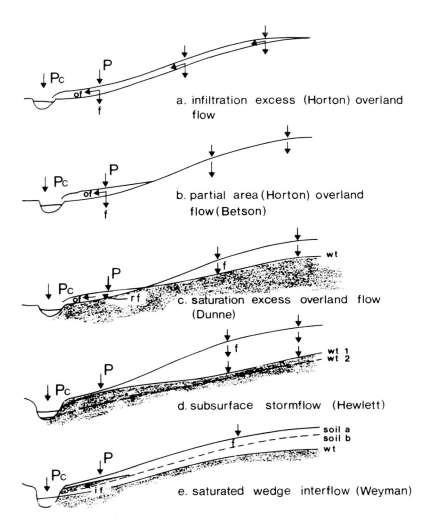

Figure 1.   Mechanisms of Runoff Production

p	precipitation
p_o	channel precipitation
f	infiltration
of	overland flow
r_f	return flow
i_f	interflow
u_f	unsaturated zone flow
c_f	channel flow
wt	water table

Mechanisms A and B, based on the infiltration capacity of the soil being exceeded by rainfall rates, continue to underlie a considerable proportion of the hydrological modelling literature. Yet, at least in vegetated humid temperate regimes, these mechanisms are at variance with several lines of evidence about catchment response. Both Kirkby (1969) and Freeze (1972) pointed out that expected infiltration rates may be high compared with natural rainfall rates and may be exceeded only rarely in humid temperate regions. The natural structure and macropores of soils in the field may increase this tendency (see discussion in Beven and Germann, 1982). The use of natural isotope data to separate discharge hydrographs into "old" or pre-event water and "new" or event water, has almost invariably resulted in high percentages of old water **at peak flow** in humid temperate areas (although few if any events greater than the mean annual flood have been included in these studies).

It is hoped that by attempting to model the dynamics of a range of runoff producing mechanisms, we may move a little further towards the goal of an acceptable physically-based model of flood frequency that may be able to explain some of the variability of flood frequency characteristics amongst catchments of different scales. This study demonstrates the use of a model that includes mechanisms A, B, C and D in a parsimonious and computationally efficient way.

3. THEORY

This study builds upon the TOPMODEL structure, variants of which have been used in the papers of Beven and Kirkby (1979), Beven and Wood (1983), Beven et al., (1984) and Beven (1986). Here, the links between the original model formulation and the hydraulic properties of the soil are made clearer, thereby also allowing the prediction of surface infiltration rates. The theory is sufficiently general to allow spatially heterogeneous soil properties to be easily incorporated.

Take any point i in a catchment. Following Beven and Kirkby (1979) we shall assume that subsurface flow rate q_i can be related to a soil storage deficit S_i by

$$q_i = K_i \, \tan\beta \, e^{-S_i/m} \qquad (1)$$

where $\tan\beta$ is the local slope angle, K_i is a soil transmissivity parameter and m is a parameter dependent on the rate of change of conductivity with depth in the profile. In many soil profiles measured saturated hydraulic conductivity K_s can be approximated by a relationship of the form

$$K_s = K_o \, e^{-fz} \qquad (2)$$

where z is depth from the soil surface, K_o is the saturated conductivity at the soil surface and f is a parameter. Beven (1984) discusses the applicability of Equation (2) and gives data showing a range of K_o from 0.01 to 100 m/hr and f from 1.0 to 13 m^{-1}. Clearly K_i in (1) may be obtained by integrating (2) from $z = 0$ to $z = D$ the lowest depth of interest, say an impermeable barrier. Then

$$K_i = (K_0/f) \, \{1 - e^{-fD}\}$$

Similarly, the subsurface flow rate $q_i \lambda$ for a water table at depth z can be obtained by integrating the conductivity function between z and d and multiplying by the slope angle $\tan\beta$. Thus:

$$q_i = (K_0/f) \{ e^{-fz} - e^{-fD} \} \tan\beta \qquad (3)$$

This is consistent with (1) only if storage deficit is distributed as a linear function of z (e.g., $S = z \, \Delta\theta_s$ where $\Delta\theta_s$ is some volumetric moisture deficit, constant with depth) and if e^{-fD} is small. The depths at which e^{-fD} is 0.1 and 0.01 for different values of f are shown in Figure 2. If this assumption holds then $K_i = K_o / f$ and $m = \Delta\phi_s / f$.

It is worth nothing here that storage deficit in this context is related to the readily drained porosity of the soil; i.e., storage above the vague limit at which vertical rates of drainage become very slow (the "field capacity" of the soil). It does not include any deficits due to evapotranspiration during dry periods. An additional root zone deficit is provided in the model to take account of such losses.

3.1 Spatially variable soil characteristics

It is now recognized that spatial variability of soil hydraulic characteristics may be important in the hydrological response of catchments. The spatial variability of K_o and K_i can be incorporated into the model in a very similar way to the treatment of topography in the original development of TOPMODEL in Beven and Kirkby (1979) if the parameter m can be assumed constant. Thus, for a steady input, r , the flow at any point i is

$$q_i = ar \tag{4}$$

where a is the area draining through i per unit contour length. Combining (1) and (4)

$$S_i = m{-}\ln(\frac{ar}{K_i \, \tan\beta}) \tag{5}$$

Integrating the point deficits over the catchment area A the mean deficit $\overline{S}$ is given by

$$\overline{S} = \frac{1}{A} \int_A S_i$$

$$= \frac{1}{A} \int_A -m \, \ln(\frac{ar}{K_i \, \tan\beta})$$

and substituting for $\ln r$ from (5)

$$\overline{S} = S_i - m\,\gamma + m \, \ln(\frac{a}{K_i \, \tan\beta})_i \tag{6}$$

where $\gamma = \frac{1}{A} \int_A \ln (\frac{a}{K_i \tan\beta})_i$ a constant for the basin. Underlying the development of (6) is an assumption that all points with the same $a / K_i \tan\beta$ value are hydrologically similar. Also the relationship (1) should hold even though $S_i < 0$, implying that return flow as well as subsurface flow are governed by (1). The use of (6) in a continuous accounting model assumes that the steady state relationships used in the development are a good approximation to storage relationships under transient conditions. If this is a reasonable assumption, then (6) allows prediction of the **pattern** of soil moisture deficit within the catchment, and particularly the saturated contributing area $(S_i \leq 0)$, from knowledge of topography and soil characteristics. Note that the basin constant γ can be separated into a topographic part and a soil part as

$$\gamma = \frac{1}{A} \int_A \ln(\frac{a}{\tan\beta}) - \frac{1}{A} \int_A \ln(K_i) \tag{7}$$

where the first integral is the topographic constant λ of Beven and Kirkby (1979).

3.2 Calculation of saturation excess surface flows

Knowledge of the pattern of deficits allows prediction of areas for which $S_i \leq 0$ at any $\bar{S}$, that is the saturated contributing area. The area for which saturation from above i.e., $(r \Delta t) > S_i - S_{uz}$ (where S_{uz} is storage above "field capacity" in the unsaturated zone) will take place in any time step can also be calculated. Total saturation excess surface runoff will be the sum of the excess water over these contributing areas. Note that all such runoff is assumed to reach a channel within one time step.

3.3 Calculation of infiltration excess runoff

The link between the spatially variable K_i values and the saturated conductivity function of Eqn. (2) allows use of the infiltration equation based on Green-Ampt assumptions presented by Beven (1984). In this model, time to ponding t_p is given by solving for the volume of infiltration at ponding $I_p = (\int t_p r(t) dt)$

$$r(t) = \frac{K_o f}{\Delta \theta} \frac{(I_p + C)}{(e^{f I_p / \theta} - 1)} \tag{8}$$

where $r(t)$ is the rainfall intensity at time t, $\Delta \theta = (\theta_s - \theta_i)$, θ_i is the initial moisture content, θ_s is the saturated moisture content and C is the storage suction factor of Morel-Seytoux and Khanji (1974).

After ponding, total infiltrated flux at any time t is given by the implicit relationship

$$t - t_p = \frac{\Delta \theta}{K_o f} \left[\ln(I + C) - \frac{1}{e^{-f C / \Delta \theta}} \{ \ln(I + C) + \sum_{n=1}^{\infty} \frac{\{-f(I + C)/\Delta \theta\}^n}{n!n} \} - \text{constant} \right] \tag{9}$$

where I is the total infiltration at time t, constant and equal to

$$I = \ln(I_p + C) - \frac{1}{e^{-f C / \Delta \theta}} \{ \ln(I_p + C) + \sum_{n=1}^{\infty} \frac{\{-f(I_p + C)/\Delta \theta\}^n}{n!n} \}$$

The model is easily adapted to time variable rainfall rates.

3.4 Calculation of subsurface contributions to streamflow

Spatial variability of K_i must also enter into the calculation of outflows from the subsurface store q_b into the channel. This will be given by

$$q_b = \frac{1}{A} \int_L K_i \tan\beta e^{-S_i / m} \tag{10}$$

where L is the channel length, the integration is taken as applying to both banks and the division by A converts q_b to a flow per unit area.

Using (6) this simplifies to

$$q_b = e^{-\gamma} e^{-\bar{S}/m} \tag{11}$$

$$= q_o \ e^{-\bar{S}/m}$$

which has the same form as the subsurface drainage equation used by Beven and Kirkby (1979).

3.5 Drainage of the Unsaturated Zone

The desire to keep the model computationally efficient particularly during interstorm periods led to restrictions being placed on the form of the equation used to predict drainage, q_v from the unsaturated zone. It was felt that rates of drainage would be closely related to the local storage deficit (indeed lag times are zero where the soil is saturated) and to the local hydraulic conductivity. The following form was chosen

$$q_v \ = \ \alpha \ K_o \ e^{-S_i/m} \tag{12}$$

i.e., α times the saturated hydraulic conductivity at the local water table, where α is the local vertical hydraulic gradient. This gradient would generally be less than 1 close to the water table but was set at 1 for the purposes of this study. Equation (12) allows analytical solutions to be obtained for changes in $\bar{S}$ during interstorm periods.

3.6 Evapotranspiration from the root zone

A single additional parameter was added to control evapotranspiration from the root zone. In keeping with studies of modelling soil moisture deficits in upland Britain (Calder et al., 1983) evapotranspiration was allowed to proceed at the potential rate until a maximum root zone deficit S_{rzmax} was reached after which evapotranspiration was set to zero. Using this procedure no separate interception calculations were necessary. Such assumptions would not however be applicable everywhere.

Potential evapotranspiration rates were modelled using the simple sinusoidal function that Calder et al. (1983) have shown gives good results in modelling measured soil moisture deficits for a range of upland and lowland sites in England. This was generalized to

$$E_p(t) = \bar{E}_p(1+\sin(0.0174t - b)) \tag{13}$$

where $\bar{E}_p$ is the mean annual potential evapotranspiration rate, t is time in days and b is a coefficient approximately equal to $\pi/2$.

4. OPERATION OF THE MODEL

The simplified theory presented above requires that subsurface storage be represented as a sequence of successive steady states. This implies that the time step of the soil moisture accounting should be long enough for the effects of upslope inputs to be transmitted to the channel bank. On the other hand the time step should not be so long that the dynamics of the variable contributing area within storm period cannot be simulated. This study has used hourly accounting periods.

Flux and water balance calculations are made for a number of "points" in the catchment. These "points" differ in their associated value of $a/\tan\beta$ and K_i. If it can be assumed that infiltration excess surface runoff is unlikely to occur then calculations can be made for a number of points of different $(a/K_i \tan\beta)$ value leading to a very computationally efficient way of taking account of spatially variable topography and soil characteristics. It is necessary

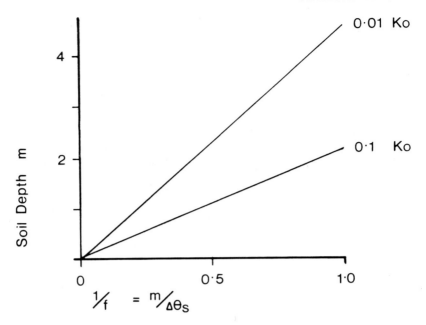

Figure 2.   The depth at which saturated hydraulic conductivity is reduced to 0.1 and
0.01 of its surface value for difference values of 1/f

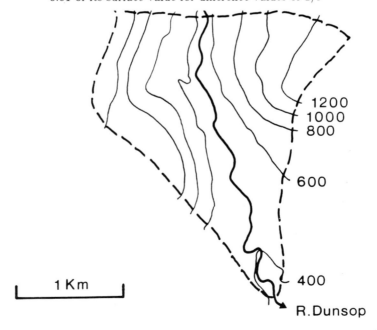

Figure 3a.   Catchment Map of Lower Dunsop (contours in feet)

to use only sufficient number of points to characterize the distribution function of $(a/K_i \tan\beta)$ for the catchment, in a similar way to the original model which assumed a homogeneous soil.

In this study, however, the possibility of infiltration excess production was not precluded. Surface infiltration rates are assumed to depend only on the local value of K_o and not $(a/\tan\beta)$, whereas local soil storage deficits depend on both. Consequently calculations are performed on a two dimensional array of "points" or "accounting cells" differing in K_o and $(a/\tan\beta)$. Each accounting cell is associated with a joint probability of occurrence of the specific values of K_o and $(a/\tan\beta)$ in the catchment. These probabilities sum over all the accounting cells to unity and are used to weight the contribution of each accounting cell to the overall catchment water balance. In this study 50 accounting cells have been used in each simulation, formed from 5 K_o values and 10 $(a/\tan\beta)$ values. In this way, the diversity of hydrological conditions within a catchment may be represented.

Clearly, actual patterns of soil hydraulic conductivity can be taken into account, if the knowledge were available, but Beven (1983) has suggested that even in very small catchments the actual pattern of hydraulic conductivities makes little difference to the occurrence of predicted saturation excess runoff, it those conductivities conform to a spatially stationary distribution (the non-uniform homogeneous case of Greenkorn and Kessler, 1969). This differs from the conclusions of Smith and Hebbert (1979) who looked at infiltration excess runoff at the single hillslope scale taking account of "run-on" infiltration. Here, it has been assumed that the probability of occurrence of the K_o values does not vary with $(a/\tan\beta)$. A log normal distribution of K_o has been used with a logarithmic standard deviation of 1.0. This is in keeping with studies of the spatial variability of saturated hydraulic conductivity (e.g., Russo and Bresler, 1982; Vieira et al., 1981).

4.1 Channel Routing

In the small catchments that have been simulated in this study, channel routing has not had an important control on the form of the storm hydrograph. A simple routing procedure based on a constant channel wave velocity and a network link length histogram has been used. There is some field evidence for the use of a constant wave velocity and linear routing procedure in catchments of this type (Beven, 1979). Geomorphological unit hydrograph techniques have not been pursued because we do not agree with generalizing the network in making predictions for specific catchments of larger scale. Surkan (1969) has shown how topologically similar networks may produce different impulse-response functions. Since the form of the channel network is one of the easiest pieces of information to obtain about a catchment there appears to be no reason why it should not be used directly in a routing procedure.

4.2 Physically-based flood frequency prediction

The requirements for a physically-based flood frequency model have been outlined by Eagleson (1972) and include: a model of the rainfall process; a model of the runoff production process; and a methodology for deriving the frequency distributions given the models of rainfall and runoff. One earlier study (Cordova and Rodriguéz-Iturbe, 1983) has used recorded maximum rainfalls of different durations to derive the frequency distribution of the annual maximum flood by simulation. Other recent studies cited above have used the method of derived distributions to obtain the flood frequency distribution from specified distributions of rainstorm intensity and duration.

Due to the complexity of the model of runoff production outlined above, it has not proved feasible to use the method of derived distributions in this study. Instead Monte Carlo simulation has been used to generate rainstorms from specified distributions of intensity and duration. Following Eagleson (1972), exponential distributions have been used, with the

Integrating (15) gives $\bar{S}$ at the end of the interstorm period as

$$\bar{S} = m \ \ln\{e^{\bar{S}_o/m} + \frac{(q_o - C_{uz})t}{m}\}$$
(16)

where $\bar{S}_o$ is the value of $\bar{S}$ at the start of drainage and t is the time to the next storm from the start of drainage.

Equation (16) holds only as long as unsaturated zone storage remains positive in each of the accounting cells contributing to $\sum_i q_v$. The value of the constant C_{uz} will change every time $S_{uz,i}$ in a particular cell goes to zero. For simplicity here C_{uz} has been held constant in each interstorm period at its value at the start of drainage and the drainage equation applied to the total unsaturated zone equation as

$$\frac{dS_{uz}}{dt} = - C_{uz} \, e^{-\bar{S}/m}$$
(17)

where $S_{uz} = \sum_i \rho_i \ S_{uz,i}$.

The time at which S_{uz} reaches zero is obtained by integrating (17) given equation (16) as

$$t_{S_{uz}} = \frac{m e^{\bar{S}_o/m}}{(q_o - C_{uz})} \exp - \{S_{uz} \frac{(q_o - C_{uz})}{m C_{uz}} - 1\}$$
(18)

If this occurs within the interstorm period (16) must be applied for the remaining time with $C_{uz} = 0$. If not, any remaining unsaturated zone storage at the start of the next storm is redistributed as a constant proportion of the initial value in each cell at the start of drainage. The coding error in the interstorm calculations noted in Beven (1986) has been corrected here but was found to make very little difference to the results.

For the root zone, evapotranspiration continues at the potential rate unless a maximum root zone deficit is exceeded. Total potential evapotranspiration during the period of drainage is given by integrating Eqn. (13) over the appropriate time interval.

5. APPLICATION OF THE MODEL

Initial runs of the model have been used to investigate

(i)   the influence of assumed soil properties on predicted mechanisms of catchment response;

(ii)   differences in predicted flood frequency characteristics in different subcatchments of a larger catchment.

Topographic data have been taken from 1:25000 scale maps of the R. Hodder catchment (256 km^2) in Lancashire, England. Three subcatchments have been investigated (Figure 3): Bottoms Beck (11.1 km^2), the Upper Loud (14.25 km^2), and the Lower Dunsop (4.3 km^2). These subcatchments were chosen to represent the extremes of topography within the Hodder catchment. Distributions of $\ln(a/\tan\beta)$ were derived for each subcatchment, by dividing the hillslopes into plane segments of constant slope but variable width, and digitizing the resulting discretization. The resulting topographic distribution functions are shown in Figure 4.

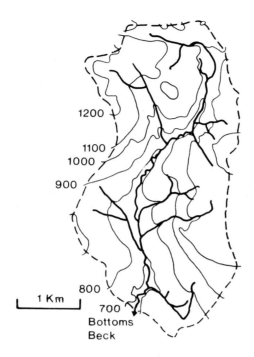

Figure 3c.   Catchment Map of Bottoms Beck (contours in feet)

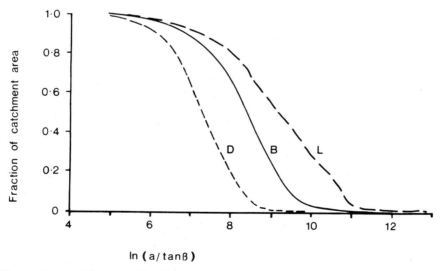

Figure 4.   Topographic distribution functions for the three catchments
L Upper Loud
B Bottoms Beck
D Lower Dunsop

mean $\ln(K_o)$ in each subcatchment. Although the model is computationally efficient compared with more detailed physically-based models, the computational burden of the accounting cell technique remains quite high. Simulations were therefore limited to 100 years (approximately 10,000 rainstorm events) but were made more directly comparable by using the same randomly chosen sequence of events for each simulation. Summary hydrological data for one of the simulations is given in Table 1, and would be representative of conditions in western upland Britain.

TABLE 1. Hydrological data for a typical 100 year
simulation (m = 0.01, m,K = 0.3 m/hr)

No. of storms	11688	
Mean storm rainfall	17.5	mm
Mean storm intensity	2.7	mm/hr
Mean interstorm duration	6.51	hr
Mean interstorm duration	68.47	hr
Mean annual rainfall	2044.7	mm/hr
Mean annual discharge	1501.6	mm/yr
Mean annual actual	541.9	mm/hr
evapotranspiration		

Values of all the parameters used in the model are given in Table 2, and of the catchment constants γ and q_o in Table 3. Simulations were carried out for 2 values of m and three sets of hydraulic conductivity distributions.

TABLE 2. Parameter Values Used in the Model

m		0.01	0.25	m
mean K_o	30	0.3	0.03	m/hr
$\sigma\ln(K_o)$		1.0		
Maximum root zone storage		0.1		m
Channel routing velocity		3600		m/hr
$\Delta\theta_s$		0.2		
Storage-suction factor C		0.04		m

6. RESULTS

The results of the simulations are summarized in Tables 4 and 5. All the distributions of peak flows greater than the threshold used to reduce the number of peaks to be analyzed were exponential in form (e.g., Figure 5a), with a Gumbel or extreme value type 1 distribution for

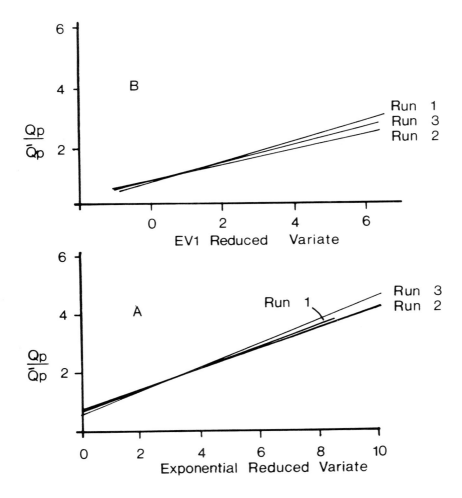

Figure 5.   A. Exponential distribution plot of all peaks over a threshold of 5 mm/hr
B. Gumbel distribution plot of maximum annual peaks for the 100 years of record simulated

TABLE 3. Catchment Constants Used in the Model

		mean K_o m/hr	γ hr/m
Bottoms Beck	30	7.13	79.8
		0.3	1.74
		0.03	16.34
Upper Loud	30	8.66	17.24
		0.3	13.27
		0.03	17.87
Lower Dunsop	30	6.85	106.21
		0.3	11.45
		0.03	16.05

the annual maximum flood peaks (e.g., Figure 5b). Beven (1985) in applying a similar model to the Wye catchment found the same result, although the plot of observed **instantaneous** annual peaks showed a tendency towards an Extreme Value Type 2 distribution. This difference may be in part due to the time lumping inherent in the model calculations for a small catchment, and perhaps also due to the rainfall model used to drive the simulations.

For the simulations with m = 0.01 m, the mean peak flow for all storms over the threshold was surprisingly consistent over all the catchments and conductivity distributions, despite an increase in the number of peak flows with decreasing conductivity. Mean annual flood increased with decreasing conductivity although the slope of the growth curve became less steep (Figure 5b). Predicted surface runoff coefficients and contributing areas also increase with decreasing hydraulic conductivity. Surface runoff dominates the runoff response at peak flows for all the simulations even with very high hydraulic conductivities. This surface runoff is predicted as being produced by a saturation excess mechanism only, for all but the lowest conductivity distribution used and even then only a fraction of the catchment is producing infiltration excess runoff whereas nearly all the catchment is producing saturation excess runoff. Some histograms of the maximum storm contributing areas generated by the model are shown in Figure 6, demonstrating the range of behaviors predicted by the model.

The different catchments demonstrate quite similar behavior especially in the form of their mean annual flood growth curves (e.g., Figure 7a).

Contributing areas and surface runoff percentages are highest in the Upper Loud catchment which has lower slope angles, but are high in all the catchments for these flood events. The Lower Dunsop "sideslope" catchment shows the most dramatic rise in the number of peaks greater than the threshold and the highest mean annual flood. At first this appears surprising but this catchment has the most steeply rising topographic distribution function (Figure 4). Thus, at low flows, there will be little contributing area, and more water enters the soil.

Although q_o values are higher for this catchment area (Table 3) the higher drainage rates do not appear to compensate for the more rapid spread of the contributing area under wet

TABLE 4. Results of 100 year simulations with m = 10 mm and different values of mean ln(K_0) for all storm peaks over a threshold of 5 mm/hr. Standard deviations are given in brackets.

Catchment	Simula-tion No.	Mean K_o	No. of Storm Peaks	Mean Peak Flow Threshold	Mean Annual Flood
		m/hr		mm/hr	mm/hr
Bottoms Beck	1	30	240	7.92(3.41)	9.94(4.28)
	2	0.3	1509	7.82(3.21)	15.26(5.00)
	3	0.003	2075	7.99(3.27)	16.74(5.04)
Upper Loud	4	30	424	7.61(3.08)	10.64(4.33)
	5	0.3	1509	7.70(3.05)	14.84(4.61)
	6	0.003	1759	7.83(3.07)	15.6(4.65)
Lower Dunsop	7	30	272	8.03(3.66)	10.44(4.81)
	8	0.3	2122	8.32(3.75)	18.56(5.88)
	9	0.003	2805	8.59(3.84)	20.18(5.92)

Catchment	Simula-tion No.	Surface Runoff/ Total storm Runoff	Surface Runoff at Peak	Saturation Excess Contributing Area	Infiltration Excess Contributing Area
		%	%	%	%
Bottoms Beck	1	56.9(8.60)	85.7(5.3)	77.6(8.7)	0.0(0.0)
	2	91.2(6.2)	99.5(1.0)	92.9(3.7)	0.0(0.0)
	3	99.4(1.2)	99.9(1.0)	99.8(0.2)	17.21(19.1)
Upper Loud	4	72.3(8.9)	96.3(1.4)	79.9(6.4)	0.0(0.0)
	5	95.8(3.5)	99.7(1.0)	95.7(2.5)	0.0(0.0)
	6	99.7(0.3)	99.9(0.0)	99.7(0.2)	19.7(19.6)
Lower Dunsop	7	54.0(9.4)	85.0(6.1)	78.1(8.8)	0.0(0.0)
	8	89.5(7.6)	99.6(1.0)	92.4(4.5)	0.0(0.0)
	9	99.4(0.7)	99.9(0.0)	99.6(0.2)	13.2(17.9)

TABLE 5. Results of 100 year simulations with m = 25 mm and
different values of mean (K_0) for Storm peaks over a threshold
of 5 mm/hr (except for simulations marked * when 2 mm/hr used).
Standard deviations are given in brackets.

Catchment	Simulation No.	Mean K_o	No. of Storm Peaks	Mean Peak Flow Threshold	Mean Annual Flood
		m/hr		mm/hr	mm/hr
Bottoms Beck	10	30*	260	3.69(2.35)	5.15(3.08)
	11	0.3	1110	7.48(2.90)	13.2(4.47)
	12	0.003	2033	7.96(3.24)	16.57(5.03)
Upper Loud	13	30*	1170	3.35(1.76)	6.95(3.22)
	14	0.3	1332	7.54(2.87)	13.86(4.32)
	15	0.003	1743	7.82(3.09)	15.54(4.64)
Lower Dunsop	15	30*	223	4.01(2.67)	5.46(3.30)
	17	0.3	1628	7.85(3.32)	15.89(5.25)
	18	0.003	2757	8.57(3.83)	20.04(5.89)

	Simulation No.	Surface Runoff/ Total storm Runoff	Surface Runoff at Peak	Saturation Excess Contributing Area	Infiltration Excess Contributing Area
		%	%	%	%
Bottoms Beck	10	5.15(3.08)	73.4(8.7)	39.0(11.2)	0.0(0.0)
	11	86.3(9.0)	99.2(1.0)	80.6(5.7)	0.0(0.0)
	12	99.0(1.5)	99.9(1.0)	98.4(1.2)	17.8(18.3)
Upper Loud	13	30.5(10.8)	73.4(8.7)	39.0(11.2)	0.0(0.0)
	14	93.7(4.97)	99.5(1.0)	88.8(2.5)	0.0(0.0)
	15	99.4(0.7)	99.9(0.0)	99.5(0.2)	19.3(18.8)
Lower Dunsop	16	30.5(10.8)	73.4(8.7)	39.0(11.2)	0.0(0.0)
	17	83.6(1.1)	99.2(1.1)	80.0(5.9)	0.0(0.0)
	18	98.7(1.6)	99.9(0.0)	99.1(0.6)	13.2(17.3)

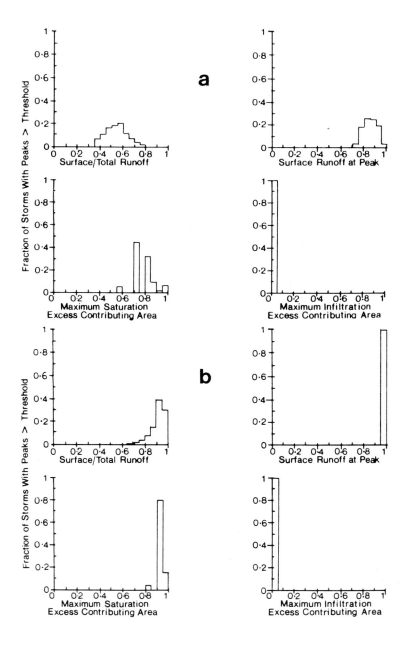

Figure 6.   Percentage surface runoff at peak, maximum saturation
excess contributing area and maximum infiltration
excess contributing area, for all storm peaks over the
specified threshold

(a) Run 7; (b) Run 2; (c) Run 3; (d) Run 13; (e) Run 17

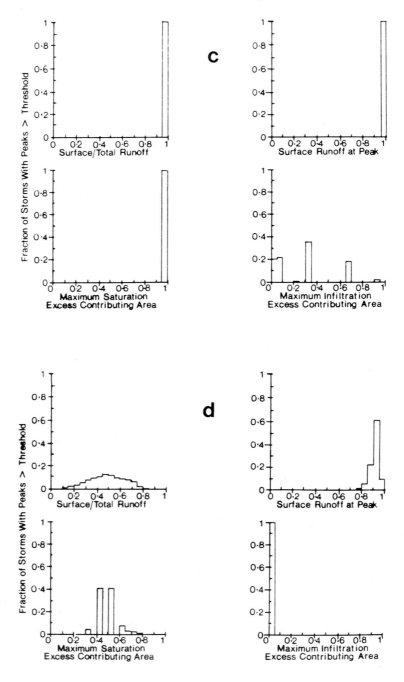

Figure 6. (con't)   Percentage surface runoff at peak, maximum saturation excess contributing area and maximum infiltration excess contributing area, for all storm peaks over the specified threshold

(a) Run 7; (b) Run 2; (c) Run 3; (d) Run 13; (e) Run 17

conditions implied by the steep topographic distribution function.

For the m = 0.025 m simulations (implying deeper soils with greater lateral flow capacities) the results show that for the highest conductivities contributing areas and peak flows are much lower, whereas for the lowest conductivities results are very similar with saturation excess contributing areas close to 100%. Infiltration excess contributing areas for the low conductivity cases are somewhat lower than with m = 0.01 due to the less steep drop in conductivity with depth.

Annual flood growth curves are slightly steeper for the high conductivity case than the m = 0.01 simulations, but very similar for the lower conductivities (Figure 7b). The differences between the catchments are similar to the m = 0.01 simulations.

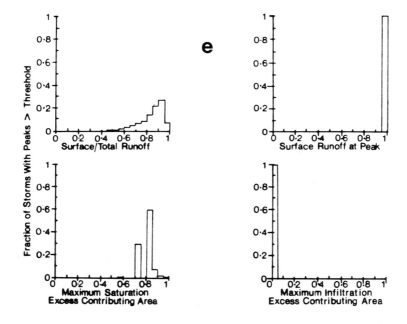

Figure 6. (con't)   Percentage surface runoff at peak, maximum saturation
                        excess contributing area and maximum infiltration
                        excess contributing area, for all stirm peaks over the
                        specified threshold

                        (a) Run 7; (b) Run 2; (c) Run 3; (d) Run 13; (e) Run 17

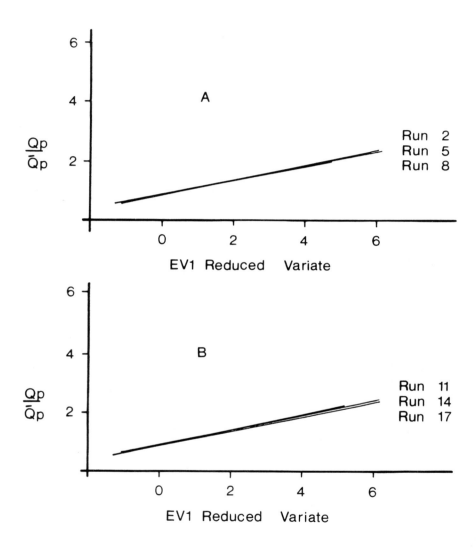

Figure 7.   Gumbel plots of maximum annual floods for each catchment with
$\overline{K}_o = 0.3m/hr$

(A) m = 0.01 m
(B) m = 0.025 m

7. CONCLUSIONS

Perhaps the most remarkable aspect of the study has been the similarities in the flood response of the different catchments over a wide range of hydraulic conductivity. Because of the dominance of saturation excess surface flow in the flood response mean annual flood rises by a factor of only 3 with a change in mean conductivity of 4 orders of magnitude. The annual flood growth curves are also remarkably consistent.

Freeze (1980) concluded from his study of runoff production on a hypothetically straight hillslope that:

> "The simulations carried out in this study have placed the author in some awe of the delicate hydrologic balance on a hillslope. If one fixes the mean hydraulic conductivity of a hillslope, then there is only a very narrow range of topographic slopes that can lead to runoff generated by the Dunne [saturation excess] mechanism. If one fixes the topographic slope then there is only a very narrow range of hydraulic conductivities that will lead to a water table that is high enough to allows the Dunne mechanism to be operative in a given climatic regime. The fact that the Dunne mechanism is so common in nature ... infers a very close relationship between climate, hydraulic conductivity and the development of geomorphic landforms." (p. 406).

In fact this study suggests that for the more realistic topographies considered here and the wet conditions associated with flood flows, it requires exceedingly high hydraulic conductivities **not** to generate saturation excess surface runoff. It is clearly difficult to avoid surface saturation in shedding surplus water over a long period of time. This conclusion is, of course, dependent on the particular parameter values used which imply soils with a relatively shallow hydrologically effective layer. The hillslopes are also quite long in this area. In the case of the Lower Dunsop catchments in particular this will compensate to some extent for the steep slope angles and lack of convergence in generating surface saturation under wet conditions. It should also be noted that an isotropic soil has been assumed. In some soils lateral downslope hydraulic conductivities may differ from vertical conductivities. It is the lateral transmissivity that is crucial in the occurrence of soil saturated to the surface under flood conditions.

The high values of contributing area and similarity in growth curves suggest that the rainfall model may have a very significant control of the flood frequency characteristics predicted by the model. Correctly characterizing the rainfall inputs and predicting runoff production are the keys to physically-based predictions of the flood frequency characteristics of catchments. The model presented here makes it possible to simulate long periods of time in a computationally efficient way. The model predicts a variety of runoff production mechanisms and the way they are affected by heterogeneity of topography and soil characteristics.

It is intended to extend this study to catchments of larger scale taking into account the spatial variation of rainfalls, routing through the actual channel network and variations in soil type.

ACKNOWLEDGMENT

This work has been funded by the Ministry of Agriculture, Fisheries and Food, UK.

REFERENCES

Beven, K.J., 1979, On the generalised kinematic routing method, *Water Resources Research*, 15(5), 1238-1242.

Beven, K.J., 1983, Introducing spatial variability into TOPMODEL: theory and preliminary results. Report to Department of Environmental Sciences, University of Virginia.

Beven, K.J., 1984, Infiltration into a class of vertically non-uniform soils, *Hydrological Sciences J.*, 29(4), in press.

Beven, K.J., 1985, Towards the use of catchment geomorphology in flood frequency predictions, *Earth Surface Processes and Landforms*, in press.

Beven, K.J., 1986, Hillslope runoff processes and flood frequency characteristics, in A.D. Abrahams (ed.), *Hillslope Processes*, George, Allen and Unwin, in press.

Beven, K.J. and Germann, P.F., Macropores and Water Flow in Soils, *Water Resources Research*, 18(5), 1311-1325.

Beven, K.J. and Kirkby, M.J., 1979, A physically-based variable contributing area model of basin hydrology, *Hydrological Sciences Bull.*, 24(1), 43-69.

Beven, K.J., Kirkby, M.J., Schoffield, N. and Tagg, A., 1984, Testing a physically-based flood forecasting model (TOPMODEL) for three UK catchments, *J. Hydrology*, 69, 119-143.

Beven, K.J. and Wood, E.F., 1983, Catchment geomorphology and the dynamics of runoff contributing areas, *J. Hydrology*, 65, 139-158.

Calder, I.R., Harding, R.J. and Rosier, P.T.W., 1983, An objective assessment of soil moisture deficit models, *J. Hydrology*, 60, 329-355.

Cordova, J.R. and Rodríguez-Iturbe, I., 1983, Geomorphologic estimation of extreme flow probabilities, *J. Hydrology*, 65, 159-173.

Diaz-Granados, M.A., Valdés, J.B., and Bras, R.L., 1984, A physically-based flood frequency distribution, *Water Resources Research*, 20(7), 995-1002.

Dunne, T., 1978, Field studies of Hillslope Flow Processes, in M. Kirkby, (Ed), *Hillslope Hydrology*, Wiley, Chichester, 227-294.

Eagleson, P.S., 1972, Dynamics of flood frequency, *Water Resources Research*, 8(4), 878-898.

Freeze, R.A., 1972, Role of subsurface flow in generating surface runoff. 2 Upstream source areas, *Water Resources Research*, 8(5), 1272-1283.

Freeze, R.A., 1980, A stochastic conceptual analysis of rainfall- runoff processes on a hillslope, *Water Resources Research*, 16(2), 391-408.

Greenkorn, R.A. and D.P. Kessler, 1969, Dispersion in heterogeneous nonuniform anisotropic porous media, *Ind. Eng. Chem.*, 61(9), 14-32.

Hebson, C. and Wood, E.F., 1982, A derived flood frequency distribution, *Water Resources Research*, 18(5), 1509-1518.

Kirkby, M.J., 1969, Infiltration, throughflow and overland flow in Chorley, R.J., (Ed.), *Water, Earth and Man*, Methuen, London, 215-227.

Morel-Seytoux, H.J. and Khanji, J., 1974, Derivation of an equation of infiltration, *Water Resources Research*, 10, 795-890.

NERC, 1975, *Flood Studies Report* (5 volumes) Natural Environment Research Council, Institute of Hydrology, Wallingford, Oxon, UK.

Rodríguez-Iturbe, I. and Valdés, J.B., 1979, The geomorphological structure of hydrologic response, *Water Resources Research*, 15(6), 1409-1420.

Rodríguez-Iturbe, I., Gonzalez, M. and Bras, R.L., 1982, A geomorphoclimatic theory of the instantaneous unit hydrograph, *Water Resources Research* 18(4), 877-886.

Russo, D. and Bresler, E., 1982, Soil hydraulic properties as stochastic processes: II Errors of estimates in a heterogeneous field, *Soil Sci. Soc. Amer. J.*, 46, 20-26.

Smith, R.E. and Hebbert, R.H.B., 1979, A Monte Carlo analysis of the hydrologic effects of spatial variability of infiltration, *Water Resources Research*, 15(2), 419-429.

Surkan, A.J., 1969, Synthetic hydrographs: Effects of Network Geometry, *Water Resources Research*, 5, 112-128.

Vieira, S.R., Nielsen, D.R. and Biggar, J.W., 1981, Spatial variability of field-measured infiltration rates, *Soil Sci. Soc. Amer. J.*, 45, 1040-1048.

7

A STUDY OF SCALE EFFECTS IN FLOOD FREQUENCY RESPONSE

Charles S. Hebson
Eric F. Wood

ABSTRACT

The effects of relative climatic and catchment scales on flood frequency response are studied with the aid of a dimensionless derived flood frequency equation. The dimensionless frequency is developed by applying the method of derived distributions from probability theory to an Instantaneous Unit Hydrograph (IUH) runoff model and a probabilistic areal rainfall model. The derived distribution approach provides a theoretical framework for treating the scale interactions in a systematic way, while the dimensionless formulation makes for a straightforward generalization of the results. The component models in the frequency are characterized by various scales, so that the frequency itself is an implicit function of those scale effects. Catchment/climate interaction works in two notable ways: 1) through the ratio λ^* of characteristic storm duration to catchment response time and 2) through the shape of the input areal rainfall intensity distribution as it is affected by the relative correlation and catchment scales, lumped in the dimensionless correlation parameter b^*. With regards to catchment scale, these two parameters have opposite effects on frequency skewness. However, in most cases the areal averaging effect is dominant and the net effect shows that flood frequency behavior in small catchments should be flashier and more highly skewed than in large catchments. These same properties are often observed in real data.

1. INTRODUCTION

The significance of scale effects in hydrology has been recognized for some time now, though systematic analysis of those effects has only recently been taken up by investigators. Attention has increasingly turned to the analysis of hydrologic response at the basin scale but has concentrated mainly on roles of catchment size and structure. The driving climatic forces on a basin are also subject to various scale factors which in turn must influence the catchment response. These interactions of catchment and climatic scale effects will be examined with respect to catchment flood frequency behavior: how are the interactions between catchment

133

V. K. Gupta et al. (eds.), Scale Problems in Hydrology, 133–158.
© *1986 by D. Reidel Publishing Company.*

and climatic scales manifested in the occurrence of extreme flow events?

A useful starting point in the analysis is to posit all random behavior in the climatic inputs that drive catchment runoff. The catchment is assigned the deterministic role of transforming random climatic behavior into the randomness observed in flood series. This working hypothesis is the basis of simulation studies of flood frequency (e.g., Ott, 1971; Leclerc and Schaake, 1972; Wood and Harley, 1975) and of runoff production (e.g., Freeze, 1980).

While physically-based simulation models can be related to the relevant scale quantities, as a practical matter they are not appropriate for tracing out the general influences at work. The effort involved in running simulations for a large number of catchment-climate configurations would be enormous and the interpretation of the results would be problematic at best. Instead, the simulation approach is better suited to specific situations where an accurate, detailed model of the process is essential. Analytical methods are better suited to a study of similarity and regional flood frequency.

The analytical analog to simulation is the method of derived distributions from probability theory. It is a mathematical technique for deriving the distribution of a function of a random variable, providing the function is simple enough. Therefore, derived distributions will be used to form a runoff distribution from an Instantaneous Unit Hydrograph (IUH) and a probabilistic rainfall model. The great advantage of this approach is that an analytical or flood frequency equation is obtained. This equation can be evaluated for a wide variety of catchment-climate scenarios with a minimum of effort. This is in contrast to the inordinate computational burden of a simulation study. Furthermore, if the component runoff and rainfall models are formulated in terms of physically meaningful parameters, then the flood frequency behavior can be related to observable catchment and climatic properties. Thus, the implicit dependence of flood frequency response on the various catchment and climatic scales is readily established if suitable scale-dependent formulations of the component models are employed. Finally, the full power of the derived distribution approach can be realized if it is reduced to a dimensionless form. Then a dimensionless flood frequency equation is obtained which is broadly applicable to a wide variety of physical situations. The importance of the dimensionless flood frequency analysis is that seemingly different physical catchments and climates will produce the same dimensionless flood frequency curve. This dimensionless analysis constitutes a useful framework for the study of hydrologic similarity with respect to flood frequency response.

The conceptual framework of a lumped IUH runoff model operating on areal rainfall inputs is appropriate in the present context and is not without technical merits in its own right. It is easy to visualize different scale effects operating within this simple model. The catchment response will surely vary with catchment size, as numerous investigators have demonstrated, with regards to basin response time. For example Gray and Wigham (1973) have summarized a variety of formulae for lag time and time of concentration for various basins.

The areal rainfall inputs are dependent on several different interacting scale influences. Point rainfall is the basic observable process from which areal rainfall must be inferred. Assuming a homogeneous and isotropic point rainfall process, the areal rainfall is seen to be a function of the magnitude of the point process, the spatial correlation of the point process, and the size of the averaging area. Taking the term "scale" in the most general sense, then, the areal average rainfall is influenced by the interaction of three distinct scales: 1) magnitude of the point process; 2) areal extent of point correlation; 3) size of averaging area. This third scale effect is essentially identical to the catchment size effect that acts on the catchment response.

The derived frequency can now be cast directly in terms of the component runoff and rainfall models in analytic or semi-analytic form. If these component models can be parameterized in terms of the scale quantities, then the ultimate effects of catchment and climatic scale on flood frequency response are readily determined. A full exploitation of the derived distribution approach therefore depends on the availability of suitable models. Fortunately, a useful runoff model is available in the form of the Geomorphologic Unit Hydrograph (GUH) (Rodriguéz-Iturbe and Valdés, 1979), an IUH derived from conceptual/theoretical considerations of basin structure. Hebson and Wood (1982) derived a flood frequency from the GUH and independent exponential distributions of rainfall intensity and duration. That work is expanded upon here in three ways: 1) the assumption of exponential distributions has been relaxed to allow for any independent distributions; 2) the explicit dependence on the GUH has been relaxed to allow for any IUH formulation; 3) the derived frequency has been recast in a dimensionless form which is readily "scaled up" to the properties of a specific physical catchment-climate configuration. All of these improvements enhance the suitability of the derived frequency as a device for studying the related issues of scale effects and similarity in flood frequency response.

The application of derived distributions to hydrologic analysis was initiated by Eagleson (1972) in his derivation of a flood frequency distribution from an approximate kinematic wave model and independent exponentially distributed climatic inputs. Wood (1976) utilized a general form of Eagleson's (1972) formulation to assess the effects of model parameter uncertainty. Eagleson (1978) expanded greatly upon his initial efforts when he analytically determined the probabilistic structure of annual water balance dynamics. Chan and Bras (1979) studied the urban runoff problem within the context of derived distributions. Cordova and Bras (1981) utilized the basin method in their formulation of a probabilistic infiltration model. Rodriguéz-Iturbe et al., (1982) formulated probability distributions for the peak and time to peak of the GUH using derived distributions. Around the same time, Hebson and Wood (1982) presented their initial derived frequency based on the GUH and exponential distributions of rainfall intensity and duration. More recently, Dias-Granados et al., (1984) derived a flood frequency model using the GUH and the physically-based Philip (1957) infiltration equation. While the great majority of applications of derived distribution methods have been to rainfall-runoff problems, the stormwater quality application of Loganathan and Delleur [1984] should also be noted.

2. DERIVED FREQUENCY FOR TOTAL RUNOFF

The original flood frequency for total runoff given by Hebson and Wood [1982] was derived from a simple rainfall-runoff model based on the GUH for routing direct runoff, and independent exponential distributions of rainfall intensity and duration. The hydrologic and climatic components of the derived frequency are summarized below, followed by a listing of the generalized form of the flood frequency equations for total runoff. Details of the frequency derivation can be found in the original reference and in the development of the dimensionless frequency that follows later in the paper.

2.1. Catchment Response Model

Direct runoff resulting from excess rainfall is routed with a triangular Instantaneous Unit Hydrograph (IUH), an approximation to the smooth continuous IUH. The approximation is adequate so long as the IUH peak, u_p, and time to peak, t_p, are preserved in the triangular representation (Henderson, 1963). Given u_p and t_p, the continuity constraint (conservation of mass) can be satisfied when unit area is contained under the direct runoff IUH. It follows

$$t_k = 2/u_p$$

The equations of the linear, time-invariant triangular IUH are .

$$u(t) = \begin{cases} u_p(t/t_p), & 0 \leq t \leq t_p \\ u_p\{(t_k - t)/(t_k - t_p)\}, & t_p \leq t \leq t_k \\ 0 & t \geq t_k \end{cases}$$

where $u(t)$ is the IUH value at time t. The time integral of the IUH is commonly called the S-hydrograph and for the triangular approximation takes on a simple quadratic form:

$$s(t) = \int_0^t u(t - \tau)d\tau$$

$$= \begin{cases} t^2/t_k\, t_p, & 0 \leq t \leq t_p \\ (-t^2/t_k + 2t - t_p)/(t_k - t_p), & t_p \leq t \leq t_k \\ 1 & t \geq t_k \end{cases}$$

Even though the S-hydrograph is a function of time, $s(t)$ is itself dimensionless. The direct runoff response Q_d to an areally averaged, steady excess intensity i_e for a duration t_r is given by integrating the IUH. Storm runoff is conveniently given in terms of the S-hydrograph as

$$Q_d(t) = \begin{cases} rA_T\, i_e\, s(t), & t \leq t_r \\ rA_T\, i_e\, \{s(t) - s(t - t_r)\} & t \geq t_r \end{cases}$$

where r is that fraction of the catchment area over which runoff is generated, $0 < r < 1$. The corresponding "contributing area" A_c is just rA_T. The calculation of excess rainfall is fraught with complexity and can only be included in the derived distribution with some difficulty. Therefore, the admittedly gross simplification of lumping the whole runoff generation process into the runoff ratio r is made.

The function $U(t;t_r)$ is proportional (by $1/t_r$) to the unit hydrograph for a storm of finite duration t_r. The expression for direct runoff $Q_d(t)$ can be restated in the form

$$Q_d(t) = i_e\, U(t;t_r)$$

where

$$U(t;t_r) = \begin{cases} s(t) & T \leq t_r \\ s(t) - s(t - t_r) & t \geq t_r \end{cases}$$

Since the peak storm runoff occurs at the maximum of $U(t;t_r)$, the time to the outflow hydrograph peak T_p is easily determined by elementary mathematical optimization. For the triangular IUH, T_p is given by

$$T_p = \begin{cases} (t_k - t_p)t_r/t_k + t_p, & 0 \leq t_r \leq t_k \\ t_k & t_r \geq t_k \end{cases}$$

Evaluating the storm hydrograph $U(t;t_r)$ at the peak time T_p defines the *peak runoff function* $U_p(t_r)$. This is the dimensionless direct peak runoff that results from a steady storm of duration t_r. For the triangular IUH, $U_p(t_r)$ is evaluated as

$$U_p(t_r) = U(T_p;t_r)$$

$$= \begin{cases} (t_r/t_k)(2 - t_r/t_k), & 0 \le t_r \le t_k \\ 1, & t_r \ge t_k \end{cases}$$

Note that in this case $U_p(t_r)$ is a function of storm duration and IUH kernel length only; IUH shape is not a factor. Total runoff Q is obtained by adding a baseflow term Q_b so that

$$Q(t) = Q_d(t) + Q_b$$

2.2. Probabilistic Climatic Inputs

The rainfall duration and intensity (climatic) inputs are taken as independently distributed random variables. Real rainfall data is non-negative and often positively skewed. Both the gamma and distributions are characterized by these same traits. Eagleson (1972, 1978) reported that the gamma and exponential distributions gave an adequate fit to selected rainfall records. The exponential distribution is actually just a special case of the more general gamma distributions (and has been used in many derived hydrologic distributions) (e.g., Eagleson, 1972, 1978; Chan and Bras, 1978; Rodriguéz-Iturbe et al., 1982; Hebson and Wood, 1982; Dias-Granados et al., 1984] because of mathematical advantages relative to the other distributions.

The flood frequency used in this report is of sufficient generality so that any appropriate input distributions can be used. However, the gamma distribution has been elected for the computational examples. The two-parameter gamma distribution of the random variable x is

$$f(x) = (x/\beta)^{\alpha-1} \exp(-x/\beta)/\beta\Gamma(\alpha)$$
$$F'(x) = \text{Prob}(x < X) = \int_0^x f(x)dx = G[x/\beta,\alpha]/\Gamma(\alpha)$$

where
$f(x) =$ probability density function (pdf)
$F'(x) =$ cumulative probability function (cdf)
$\alpha =$ dimensionless shape parameter
$\beta =$ scale parameter with units of x
$\Gamma(\cdot) =$ gamma function
$G[\cdot,\cdot] =$ incomplete gamma function

The exponential distribution corresponds to $\alpha = 1$. The exceedance probability $\text{Prob}(x > X)$ is denoted by $F(x)$. The absolute moments $m_k'(X)$ and mean, variance, and skew are, respectively,

$$m_k'(x) = \beta^k \prod_{j=1}^{k} (j + \alpha-1)$$
$$m(x) = \alpha\beta$$
$$s^2(x) = \alpha\beta^2$$
$$g(x) = 2\alpha\beta^3$$

which give coefficients of variation, $C_v = \alpha^{-0.5}$ and of skew, $C_s = 2\alpha^{-0.5}$. Note that the skew is denoted by $g(x)$ while the skew coefficient is denoted by C_s.

The shape and scale parameters for point rainfall intensity, i' , will be denoted by α' and β' , while corresponding parameters in the duration distribution will be denoted by κ and λ. Due to the simple model of a constant runoff ratio r , the distribution of point excess rainfall intensity i_e' is taken to be identical to the distribution of total point intensity. Lumped rainfall-runoff models require areally averaged rainfall inputs. The assumption is made that the durations of point and areally averaged rainfall are identical. However, the distribution of areally averaged intensity will depart significantly from the point intensity distribution due to small rainfall correlation lengths with respect to catchment scales.

Areally averaged excess intensity can be shown to follow a gamma distribution with shape parameter α and scale parameter β (Wood and Hebson, 1986).

By equating the average point rainfall excess intensity to the areal average and utilizing the relationship between the areally averaged variance and the point variance which depends upon the expected correlation between two points chosen randomly in the area, the parameters of the areal rainfall probability distribution can be related to the parameters of the point process distribution.

Various empirical procedures are also available for relating point and areal rainfall. For example, the reduction curves tabulated in various U.S. National Weather Bureau technical papers (U.S. Weather Bureau, 1958).

2.3. Method of Derived Distributions

The flood frequency distribution is derived using the Method of Derived Distributions (Lindgren, 1976), a useful technique for the mathematical formulation of the distribution of a function or transformation of a random variable. In this application, total runoff (distribution unknown) results from the rainfall-runoff transformation of random rainfall intensity and duration (distributions known). For the one-to-one transformation of the n-long random vector $\tilde{x}$ into the n-long random vector $\tilde{y}$, $\tilde{y} = \tilde{w}(x)$, the density of $\tilde{y}$ is given by the general transformation equation

$$f^{\sim}(y) = f_{\tilde{x}}(\tilde{v}(y)) \mid J \mid$$

where $v^{\sim}(y) = \tilde{x}$ is the inverse transformation and $\mid J \mid$ is the Jacobian determinant with J defined by $J = \partial \tilde{x} / \partial \tilde{y}$. The random vector $\tilde{y}$ corresponds to total runoff, the function $w(x)$ corresponds to the rainfall-runoff transformation, and the random vector $\tilde{x}$ corresponds to the climatic inputs. The derived probability density function $f^{\sim}(y)$ corresponds to the parent distribution of all direct runoff peaks, not just those extrema that comprise annual maximum and annual exceedance series. The associated flood frequency distribution for those extrema is obtained by applying some basic concepts from order statistics, a problem to be discussed shortly.

2.4. Derived Frequency for Total Runoff

The flood frequency as presented by Hebson and Wood (1982) is given in the appendix with minor modification. It was derived by applying the derived distribution transformation to the component runoff and rainfall models, all of which have been described. A detailed derivation of the expressions in the appendix is not given here as essentially the same exercise will be worked through for the more general case of a frequency for dimensionless runoff. The differences between the original derivation and the appendix can be summarized as follows: 1) the former assumptions of exponential distributions of intensity and duration have been relaxed; 2) a generic IUH has been substituted for the specific GUH formulation; 3) the original version assumed that peak storm runoff occurred at the cessation of the rainstorm; a reasonable assumption for long storms. The current analysis relaxes this assumption. Frequency calculations based on the original assumption and the correct time to peak were compared and the differences were found to be minor.

3. DERIVED FREQUENCY FOR DIMENSIONLESS RUNOFF

3.1. Introduction

A significant feature of the derived flood frequency model presented in the appendix is its ease of use when compared to alternative simulation models. By a judicious choice of component models, the flood frequency model can reflect the influence of observable catchment and climatic factors. However, the number of parameters is still quite large and generally site-specific. For practical purposes and particular applications this is not a problem. However, for similarity analyses, this trait may become a liability because it reduces the utility of the derived frequency model to that of a numerical simulation model. Conceptually it is straight-forward to evaluate the derived frequency model for a large number of parameter combinations; unfortunately an interpretation of the results would be exceedingly difficult making a complete sensitivity and similarity analysis infeasible.

Fortunately, this problem can be circumvented by casting the flood frequency model in a dimensionless form. The result is a frequency analysis for dimensionless runoff that is independent of explicit catchment and climatic scale parameters. Instead, the dimensionless frequency is governed by dimensionless parameters that are indicative of *relative* scale effects; thus it is applicable to the broadest range of physical situations that conform to the underlying model assumptions. The dimensionless flood frequency model is ideally suited to sensitivity and similarity analyses by virtue of the general and parsimonious nature of its parameterization. Furthermore, the essential behavior of the flood frequency curve can be conveyed in a concise and comprehensible manner. Compared with the original model of the appendix, it is possible to reveal the important features of the derived frequency with fewer, though more selective calculations.

This concept of a flood frequency curve for dimensionless runoff is akin to traditional dimensional analysis in hydraulics but the differences are worth noting. Dimensional analysis is based on the relevant physical variables in a process and their dimensions; thus, it is ideally suited to extending limited experimental or empirical studies to other dimensions or fluid properties. When system equations are presumed to describe the phenomenon of interest then it makes sense to fully exploit that knowledge; that is, similarity analysis can proceed on the basis of scaling the equation variables in a consistent manner resulting in dimensionless forms of the original model equations. The choice of scaling variables usually is not difficult, as characteristic values of variables in the model equations are dependable choices. The basic idea behind model scaling is simple enough and receives regular application in hydrology. For example, in much of the literature on the various kinematic wave models the equations are developed in dimensionless form (e.g., Eagleson, 1970). A thorough discussion of similarity based on model equation scaling for porous media is given by Corey (1986). In spite of the differences, though, dimensional analysis and model scaling are essentially equivalent: both seek to characterize a phenomenon in terms of a few dimensionless parameters that capture the important interactions of the different influences at work, thus establishing criteria for similarity. Criteria based on model equation scaling must ultimately be tested with real data. Depending on the model performance, a dimensional analysis might then be used in a revision of model and similarity rules.

The fundamental idea in a development of a dimensionless flood frequency model is one of normalizing the physical variables by appropriate characteristic values. The physical dimensions are thereby removed and specific physical scale efforts are lumped together in dimensionless parameters. This goal can be realized in two different ways. The flood frequency equation in the appendix can be normalized in the same way that the distribution of a linear transformation of a random variable is derived. Alternatively, the frequency can be derived from normalized component models using derived distributions. This second approach will be elected, since a familiarity with the dimensional model is of some help in interpreting the resultant dimensionless expressions.

3.2. Scaling the Component Models

Dimensionless runoff and rainfall models are derived by simple linear scaling of the various physical variables and parameters. The characteristic scaling values must be relevant to the rainfall and runoff processes and must be consistently applied to both models. There are two basic types of physical variables to consider, time (storm duration and IUH time variables) and volume flux (discharge and rainfall). Thus, a single characteristic time τ is used to normalize all time variables and a single characteristic flux Q_n is used to normalize discharge and rainfall. In the development to follow all dimensionless variables will be denoted by a superscript asterisk ("*").

Time influences the derived frequency through storm duration and the IUH. A dimensionless instantaneous unit hydrograph (DIUH) is formulated by normalizing the time variables in the triangular IUH by a characteristic time τ. Thus, the triangular DIUH is determined by the dimensionless peak u_p^*, time to peak t_p^*, and kernel length t_k^*:

$$u_p^* = u_p \tau$$
$$t_p^* = t_p / \tau$$
$$t_k^* = 2 / u_p^*$$

and the dimensionless response is calculated as

$$u^*(t^*) = \begin{cases} u_p^*(t^*/t_p^*), & 0 \le t^* \le t_p^* \\ u_p^*\{(t_k^* - t^*)/(t_k^* - t_p^*)\}, & t_p^* \le t^* \le t_k^* \\ 0, & t^* \ge t_k^* \end{cases}$$

where t^* is the normalized time, $t^* = t/\tau$. The corresponding dimensionless S-hydrograph is

$$s^*(t^*) = \begin{cases} t^{*2}/t_k^* t_p^*, & 0 \le t^* \le t_p^* \\ -(t^{*2}/t_k^* - 2t^* + t_p^*)/(t_k^* - t_p^*), & t_p^* \le t^* \le t_k^* \\ 1 & t^* \ge t_k^* \end{cases}$$

The forms of $u^*(t^*)$ and $s^*(t^*)$ do not depend on the particular choice of normalizing time τ. Similarly, the dimensionless time to peak runoff T_p^* is (Wood and Hebson, 1986).

$$T_p^* = \begin{cases} (t_k^* - t_p^*) \mid t_r \mid t_k^* + t_p^*, & 0 \le t_r^* \le t_k^* \\ t_k^* & t_r^* \ge t_k^* \end{cases}$$

and the dimensionless peak runoff function $U_p^*(t_r^*)$ is given by

$$U_p^*(t_r^*) = \begin{cases} (t_r^*/t_k^*)(2 - t_r^*/t_k^*), & 0 \le t_r^* \le t_k^* \\ 1 & t_r^* \ge t_k^* \end{cases}$$

where t_r^* is the normalized storm duration, $t_r^* = t_r / \tau$. It should be noted that the unscaled and scaled S-hydrograph functions $s(t)$ and $s^*(t^*)$ are both dimensionless and in fact are identical i.e., $s(t) = s^*(t^*)$; the same holds true for the associated functions $U(t;t_r)$ and $U_p(t_r)$.

The same time normalization must also be extended to the storm duration distribution. The distribution of t_r^* is given by

$$f(t_r^*) = f_{tr}(\tau t_r^*) dt_r / dt_r^*$$

When t_r is distributed as gamma, the distribution of t_r^* follows as

$$f(t_r^*) = (t_r^*/\lambda^*)^{\kappa-1} \exp(-t_r^*/\lambda^*)/\lambda^* \Gamma(\kappa)$$

where $\lambda^* = \lambda/\tau$. The parameter λ^* is a probability distribution scale parameter, and it represents a ratio of climatic and characteristic time factors.

Total runoff response in terms of the dimensionless response function is given by

$$Q(t^*) = rA_T i_e U^*(t^*;t_r^*) + Q_b$$

The combination of physical runoff and dimensionless time is not very useful, but the extension to dimensionless direct runoff is straightforward. Rearranging terms and dividing both sides of the equation by the normalizing flux $Q_n = rA_T \beta$ gives

$$q_d^*(t_r^*) = i_e^* U^*(t^*;t_r^*)$$

where $\quad q_d^* = (Q - Q_b)/rA_T$, normalized direct runoff
$\quad\quad\quad i_e^* = e_e/\beta$, normalized intensity.

The distribution of dimensionless intensity is especially simple, being just standard gamma, i.e., with the same shape parameter α and a scale parameter of 1:

$$f(i_e^*) = i_e^{*\alpha-1} \exp(-i_e^*)/\Gamma(\alpha)$$

The particular choice of Q_n may seem arbitrary but there are few alternatives. The nature of the problem dictates that Q_n be characteristic of the rainfall or runoff process. Since the runoff distribution is unknown, Q_n must be based on the rainfall. The mean excess intensity $(\alpha\beta)$ might have been used instead of just the scale parameter β. This approach has merit since the concept of the mean is more easily grasped than the scale parameter alone; it might also provide a better basis for comparing catchments in climates with different shape parameters. The drawback of using $Q_n = rA_T \alpha\beta$ is that the distribution of dimensionless intensity would no longer be standard gamma, a desirable simplification in its own right. The difference between the two alternatives is the minor one of linear scaling by the dimensionless parameter α.

Now that the dimensionless component models have been presented, the superscript asterisk ("*") denoting dimensionless quantities will be suppressed in the interests of a less cluttered notation. Therefore, all variables should be assumed dimensionless unless otherwise indicated. In those situations where confusion might arise, the asterisk convention will be continued.

3.3. The Derived Distribution Transformation

The derivation of the frequency distribution of all dimensionless direct runoff events follows. For a given storm duration t_r, the peak value q_p (i.e., storm hydrograph peak) is calculated as

$$q_p = w_1(i_e, t_r) = \begin{cases} i_e U_p(t_r), & t_r \leq t_k \\ i_e & t_r \geq t_k \end{cases}$$

The peak for $t_r \leq t_k$ corresponds to a transient flow momentarily realized at T_p. The other possible realization of q_p, for $t_r \geq t_k$, corresponds to a steady flow. Since two variables (i_e, t_r) transform to the single variable q_p, a second dummy variable is needed before the derived distribution transformation can be applied. The simplest approach is to use the dimensionless duration t_r, since the pairs (i_e, t_r) and (q_p, t_r) are related by a 1–1 transformation. Then the second transformation is trivial:

$$t_r = w_2(i_e, t_r) = t_r, \quad \text{for all } t_r$$

The inverse transformations follow directly:

$$i_e = v_1(q_p, t_r) = \begin{cases} q_p/U_p(t_r), & t_r \leq t_k \\ q_p & t_r \geq t_k \end{cases}$$

$$t_r = v_2(q_p, t_r) = t_r, \quad \text{for all } t_r$$

The Jacobian of the transformation is $J = \partial(i_e, t_r)/\partial(q_p, t_r)$ which results in

$$= \begin{bmatrix} 1/U_p(t_r) & -(q_p/U_p(t_r)^2)\partial U_p(t_r)/\partial t_r \\ 0 & 1 \end{bmatrix} \quad t_r \leq t_k$$

$$= \begin{bmatrix} 1 & 0 \\ 0 & 1 \end{bmatrix} \quad t_r \geq t_k$$

and the Jacobian determinant is

$$|J| = \begin{cases} 1/U_p(t_r), & t_r \leq t_k \\ 1 & t_r \geq t_k \end{cases}$$

Thus, the joint pdf of q_p and the dummy variable t_r is

$$f(q_p, t_r) = \begin{cases} f(i_e = q_p/U_p(t_r))f(t_r)/U_p(t_r), & t_r \leq t_k \\ f(i_e = q_p)f(t_r), & t_r \geq t_k \end{cases}$$

The marginal distribution of q_p is obtained by integrating over t_r. After some manipulation, the following expression is obtained:

$$f(q_p) = F(t_k)f(i_e = q_p) + L(q_p)$$

where

$$L(q_p) = \int_0^{t_k} f(t_r)f(i_e = q_p/U_p(t_r))/U_p(t_r)dt_r$$

Note that $f(q_p)$ is the weighted sum of the densities of steady and transient realizations of the flow q_p. The weights are just the probabilities of getting steady and transient flows, with Prob (steady flow) $= F(t_k)$. Under steady conditions the outflow q_p equals the input excess intensity i_e. Hence, the steady flow pdf component $f(i_e = q_p)$ is weighted by $F(t_k)$. The non-steady contribution is more complicated since the intensity required for transient q_p is a function of duration t_r; However, the reasoning is the same. An approach based on these considerations of steady and transient realizations of the specified q_p could have been used to derive the cdf $F'(q_p)$ directly.

The cdf $F'(q_p)$ is obtained by integrating the pdf:

$$F'(q_p) = \int_0^{q_p} f(q_p)dq_p$$

$$= 1 - F(t_k)F(i_e = q_p) + J(q_p)$$

where

$$J(q_p) = \int_0^{t_k} f(t_r)F(i_e = q_p/U_p(t_r))dt_r$$

The k-th absolute moment of the derived distribution is given as

$$m_k'(q_p) = \int_0^{\infty} q_p^k f(q_p)dq_p$$

$$= F(t_k)m_k'(i_e) + B_k(q_p)$$

where

$$B_k(q_p) = \int_0^{\infty} q_p^k \int_0^{t_k} f(t_r)f(i_e = q_p/U_p(t_r))/U_p(t_r)dt_r\,dq_p$$

As with the probability density function, the cumulative distribution function and absolute moments are composed of steady and transient flow components. In all cases, the steady component can be evaluated analytically since it is just a weighted value of the corresponding input intensity quantity. In most cases, though, the transient components $(L(q_p), J(q_p),$ and $B_k(q_p))$ will have to be calculated numerically.

The distribution $F'(q_p)$ is the parent distribution of all dimensionless direct runoff peaks. The distribution corresponding to the annual exceedance (AE) flood series can be derived using an approximation suggested by Eagleson (1972). Given n dimensionless direct runoff events per year, the annual exceedance probability $H_E(q_p)$ can be approximated by

$$H_E(q_p) = nF(q_p), \quad q_p \geq q_0 = F^{-1}(1/n)$$

The equivalent recurrence interval in years is

$$T_E(q_p) = 1/H_E(q_p), \quad q_p \geq q_0 = F^{-1}(1/n)$$

The condition on q_p is imposed so that recurrence intervals less than 1 year are avoided (minimum recurrence intervals besides 1 year are also possible, depending on the basic interval of interest). The pdf for the annual exceedance series distribution is

$$h_E(q_p) = nf(q_p)$$

so that the absolute moments m of the flood frequency are

$$\mu_k' (q_p) = n \{m_k' (q_p) - m_{kp}' (q_p, q_0)\}$$

Greek letters are used to denote frequency moments. The *partial moment* term $m_{kp}' (q_p, q_0)$ signifies the dimensionless runoff moment integral evaluated up to the lower limit q_0:

$$m_{kp}' (q_p, q_0) = \int_0^{q_0} q_p^k f (q_p) dq_p$$

The corresponding partial absolute moment of the gamma intensity input distribution can be evaluated with the recursion formula

$$m_{kp}' (i_e, q_0) = - q_0^{k+a-1} \exp(- q_0)/\Gamma(\alpha) + (k+\alpha-1) m_{k-1,p}' (i_e, q_0)$$

where

$$m_{0p}' (i_e, q_0) = F' (i_e = q_0) = G [q_0, \alpha]/\Gamma(\alpha)$$

Upon working out the first few intensity partial moments an explicit formula becomes apparent:

$$m_{kp}' (i_e, q_o) = \{ \prod_{j=1}^{k} (j + \alpha - 1)\} \{-(\exp(-q_0)/\Gamma(\alpha)) \sum_{j=1}^{k} q_0^{j + \alpha - 1}$$

$$/ \prod_{r=1}^{k} (r + \alpha - 1) + F' (i_e = q_0)\}$$

An exact expression for the distribution of annual maxima (AM) can be derived using a basic result from order statistics. Given the parent distribution $F' (q_p)$, the cdf $H_M (q_p)$ of the largest in a sample n-long is

$$H_M' (q_p) = F' (q_p)^n$$

with pdf $h_M (q_p)$,

$$h_M (q_p) = nF' (q_p)^{n-1} f (q_p)$$

These expressions are more difficult to work with than the annual exceedance expressions since they must be evaluated entirely by numerical means. As a consequence, the different effects of catchment and climate are not so easy to discern. Therefore, only the AE flood frequency will be considered and the indicating subscript "E" will be dispensed with. The annual maximum distribution is of some use in specific applications, though, where available data is usually in the form of annual maxima.

4. DIMENSIONLESS FREQUENCY AND RELATIVE SCALE EFFECTS

4.1. The IUH Model and Characteristic Time

The specification of a characteristic time τ has intentionally been omitted until now so that the generality of the derived distribution and scaling could be demonstrated. However, the relationship of flood frequency response to catchment and climatic scales can only be established by using an IUH model that explicitly reflects physical catchment scale properties. The choice of τ should in turn by guided by considerations of the storm duration and IUH time characteristics. Time normalization can proceed on the basis of a catchment response time or a storm duration. Some degree of flexibility is inherent in this decision. Regardless of the choice, though, after normalization one of the two component models will be devoid of all scale properties while the other will contain a lumping of IUH and duration scale effects. It was felt that the scale interaction is more easily represented in the duration distribution. Hence it was decided to normalize on the basis of a catchment response time.

The IUH can be parameterized in any number of ways. For example, the Nash conceptual linear reservoir (Nash, 1957) model is mathematically identical to a gamma probability density function (pdf) and is therefore characterized by two fitting parameters. The derivation of the equivalent triangular IUH is straightforward. Unfortunately, the parameters of the gamma IUH can only be related to catchment characteristics using empirical methods (e.g., Nash, 1960). For particular applications this approach or black-box identification method may be acceptable but it is inappropriate for the present work. A more general IUH formulation is required. Storage routing models based on a careful representation of the channel network (Boyd, 1978; Surkan, 1969) are promising but the resultant response functions are unwieldy and may still not be parameterized completely in terms of observable catchment properties. Given these limitations, the Geomorphologic Unit Hydrograph (GUH) (Rodriguéz-Iturbe and Valdés, 1979) appeared well suited for the purposes of this investigation. The GUH is based on Horton order ratios, a stream segment length scale factor, and a flow velocity dynamic factor. A full account of Horton catchment analysis and related geomorphologic concepts is given by Smart (1972). The triangular approximation to the GUH is calculated according to the following IUH characteristics:

$$u_p = 0.364R_L^{0.43} \, v \,/L_\Omega$$
$$t_p = 1.584R_{BA}^{0.55} \, R_L^{0.38} \, L_\Omega/v$$
$$t_k = 2/u_p$$

where

R_L = Horton length ratio
R_{BA} = ratio of Horton bifurcation ratio to Horton area ratio, R_B/R_A
L_Ω = length of highest order stream segment, [L]
v = flow velocity, [L/T]

The smooth, continuous form of the GUH is a more complicated expression. However, Rosso (1984) used these equations to calculate the equivalence between the simple 2-parameter gamma and GUH response models. The result was a gamma IUH given in terms of the GUH parameters.

A good choice of time constant for the triangular GUH is

$$\tau = L_\Omega/v$$

Then it follows that

$$u_p^* = 0.364R$$

$$t_p^* = 1.584R_{BA}^{0.55} \, R_L^{0.38}$$

$$t_k^* = 2/u$$

$$\lambda^* = \lambda v \,/L$$

A useful simplification that follows from these choices for IUH model and time constant τ is that the dimensionless peak runoff function $U_p^*(t_r^*)$ is a function of one parameter only, the Horton length ratio R_L. Thus, all catchment and dynamic scale effects are absent in the dimensionless GUH (DGUH) and the DGUH is a function of catchment order ratios only. The dimensionless duration scale parameter λ^* is a lumping of climate, catchment size, and flow dynamics factors, thus constituting a measure of typical storm durations relative to response time. The full power of the dimensionless frequency can now be appreciated. The effects of the parameters r, A_T, β, and Q_b in the physical flood frequency distribution are accounted for by simple scaling of the dimensionless discharge q_d^*. More importantly, the parameter λ^* captures the interaction of λ, L_Ω, and v in a manner that would be difficult to achieve in a sensitivity analysis of the original flood frequency curve. The parameters α, κ, n, and R_L by necessity remain outside the scope of the dimensional transformation. Partially because they are themselves dimensionless and furthermore, because their effects are non-linear, simple normalizations are inappropriate. A dimensionless flood frequency distribution is specific to the given parameter set $(\lambda^*, \alpha, \kappa, n, R_L)$ and we have not investigated how to relate that frequency curve to a different set.

It is easy to imagine different catchment-climate combinations giving the same λ^*. This is notable, because λ^* plays a critical role in the dimensionless flood frequency response by way of the relative weights placed on the steady and transient flow probability components of the derived cdf. Fast catchments subject to long storms will have large λ^* values and the steady flow component will dominate. In such instances, the dimensionless runoff frequency intensity will be nearly identical to the input intensity frequency. As Klemes (1978) pointed out for Eagleson's (1972) kinematic wave frequency, this is essentially the rational method (Kibler, 1982) for flood frequency determination. On the other hand, slow catchments subject to brief storms will have low λ^* values and the transient probability component will be significant. In such instances, the transformation of the intensity frequency is highly non-linear and the intensity frequency is noticeably distorted by the rainfall-runoff transformation.

The relationship between λ^* and the weighting of the steady and transient probability components is intuitively correct and also consistent with the derived frequency equations. However, a computational evaluation of the equations is imperative for a full understanding of the frequency. While the parameter λ^* is a useful tool for analyzing the climate, size, and dynamic factors, there is a hidden scale effect that is not accounted for in λ^*. One possible source of variation in λ^* is variation in L_Ω, a catchment scale parameter. As L_Ω changes, the distribution of areal rainfall should also be expected to change. This dependence is a function of the rainfall spatial correlation structure and the averaging area size.

4.2. Computational Results

The dimensionless frequency curve was tested using the geomorphologic parameters of Basin 1 in Rodriguéz-Iturbe et al. (1979b), exponential rainfall distributions, and a value of $n = 25$ storms/year. The data set is summarized in Table 1. The exponential (dimensionless) intensity frequency is the limit at high λ^* values and is given by

$$H(i_e) = nF(i_e) = n\,\exp\,(-i_e), \quad i_e > i_0 = \ln(n)$$

After some manipulation it can be shown that the intensity frequency distribution has absolute moments

$$\mu_k'(i_e) = k! + \sum_{j=1}^{k}(k!/j!)i_0^j$$

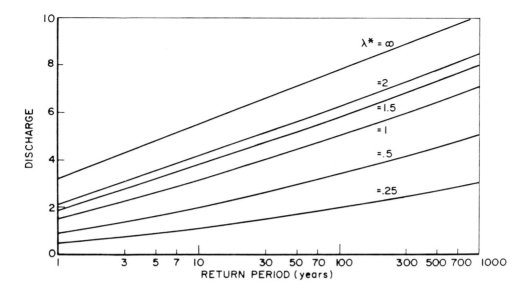

Figure 1:   Dimensionless frequency curve variation with λ^*.

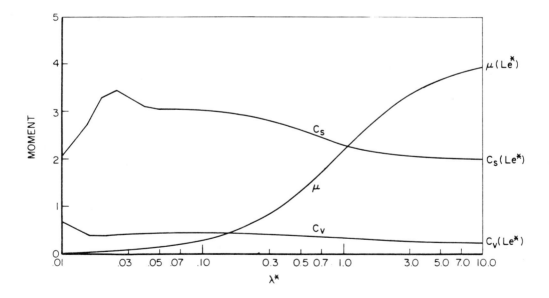

Figure 2:   Dimensionless frequency moment variation with λ^*.

In particular, the mean, variance, and skew are given by

$$\mu(i_e) = i_0 + 1$$
$$\sigma^2(i_e) = 1$$
$$\gamma(i_e) = 2$$

with

$$C_v(i_e) = 1/(i_0 + 1)$$

and

$$C_s(i_e) = 2.$$

The variation of the flood frequency with λ^* is indicated in Figure 1. Frequency curves for a range of λ^* values are shown, along with the limiting curve of $\lambda^* = \infty$ that corresponds to the rainfall intensity frequency curve. As λ^* is decreased, the transformation becomes increasingly non-linear and the frequency curves fall with respect to the intensity curve. Thus, for a given q_p, the recurrence interval T increases with decreasing λ^*, while for T held constant, q_p decreases with decreasing λ^*. In the othe. limit, λ^* tends to 0 and the frequency curve lies on the horizontal axis.

The mean and coefficients of variation and skew are plotted against λ^* in Figure 2. The respective values corresponding to the limiting case of the rainfall intensity frequency are also indicated. The mean approaches the limiting value from below while the skew and variation coefficients approach from above. Thus, the mean increases with λ^* and the skew and variation coefficients decrease with λ^*. The moment ratios are closer to the limiting values over a larger range of λ^* than the mean.

This consideration of the frequency and moments as functions of λ^* suggests that the "flashiness" of a catchment is strongly related to λ^*. Flashiness with respect to flood frequency is a relative concept, simply expressed by saying that one catchment is flashier than another if, for a given recurrence interval T, the T-year event in the first catchment is larger than in the second catchment. Alternatively, a given runoff event will have a lower recurrence interval in the flashier catchment. By these criteria, the degree of flashiness increases with λ^*. This is illustrated in Figure 3, where the 25-, 100-, 500- and 1000-year events are plotted as functions of λ^*. The same events for the intensity frequency curve are also indicated.

The analysis thus far has proceeded on the basis of a simple variation of λ^*, with everything else in the frequency held constant. However, one important cause of variation in λ^*, the length L_Ω, also necessitates an adjustment in the input areal intensity distribution for reasons already cited. As L_Ω changes, so does the catchment area and hence the distribution of areal intensity. Therefore, the above results should be interpreted as being for constant catchment size, with all variation in λ^* attributable to λ and v.

The effect of areal averaging in the dimensionless frequency was studied by simulating random intensity fields. Two catchments and two climates were considered. Catchments of orders $\Omega = 3$ and $\Omega = 5$ were used with the geomorphologic data in Table 1, resulting in a "small" catchment of 100 km^2 and a "large" catchment of 1600 km^2. Point rainfall interacts with catchment size through the spatial correlation ρ of the intensity process. A simple exponential correlation function,

$$\rho(r) = \exp(-r/b), \quad r \geq 0$$

was used, where r is the distance between two points. The correlation parameter b is a measure of the strength of the correlation process: large b corresponds to a highly-correlated process while low b corresponds to a poorly- correlated process. Note that b has units of length. The two values of b used in this example, $b_s =10$ km and $b_L =80$ km, correspond to these two extremes. The first value, $b_s =10$ km, is representative of monthly summer data in the Netherlands (Stol, 1972) while the second value, $b_L =80$ km, is more suggestive of large frontal storms. For example, a b value of approximately 200 km would be calculated for the rainfall of Hurricane Connie in the Baltimore area in August 1955 (Rodriguéz-Iturbe, and Mejia, 1974a). For such large correlation parameters, the process is nearly constant in space. Since the b parameter is indicative of the spatial uniformity of the process, the adjectives "small" and "large" will be used ot describe the climates for the respective values of $b_s =10$ km and $b_L =80$ km.

An exponential point process, $\alpha' =\beta^{*'} = 1$, was assumed as the basis of both climates, the only difference being the correlation structure. Normal (0,1) deviates were generated on a 21-point by 21-point grid using the Turning Band algorithm (Mantogloú and Wilson, 1982) according to the specified correlation structure and subsequently transformed to the probabilistically equivalent exponential deviates. While the sample correlations of the exponential fields were slightly different than for the normal fields, the minor differences were not significant in the context of this study. The Turning Bands simulations were best executed on rectangular grids. Therefore, square regions of 10 km by 10 km and 40 km by 40 km were used to mimic the small and large catchments.

TABLE 1. Parameters Used in Test of Dimensionless Frequency

PARAMETER		SPECIFICATION
area ratio R_A		4
bifurcation ratio R_B		3
length ratio R_L		1.5
1-st order area $A_1(km^2)$		6.25
1-st order length $L_1(km)$		4.58
1-st order stream number N		9
rainfall shape parameters		
	α	1
	β^*	1
number storms per year n		25

Some general comments are in order before proceeding to the simulation results. The two extreme spatial processes are characterized by absolute correlation ($b = \infty$) and no correlation (independence, $b = 0$). Now, the areal average is estimated as the simple arithmetic average of point deviates. Thus, the distribution of the areal average estimate is essentially the distribution of the sum of random variables. For the case of complete correlation, the point deviates in a single simulation are identical and the distribution of the point and areal processes are the same. For the opposite extreme, the distribution of n independent gamma (α,β) random variables is also gamma, with parameters $(n\alpha,\beta/n)$. Thus, the areal average is distributed as gamma $(n\alpha,\beta)$. For large b, the areal process should not be much different

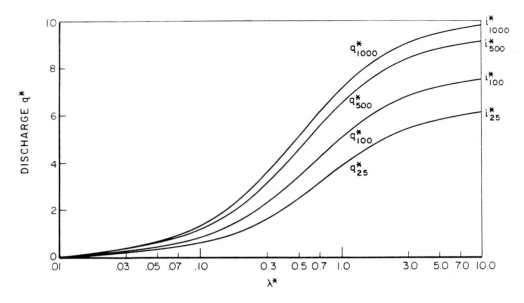

Figure 3:   Variation of T-year event with λ^{*}.

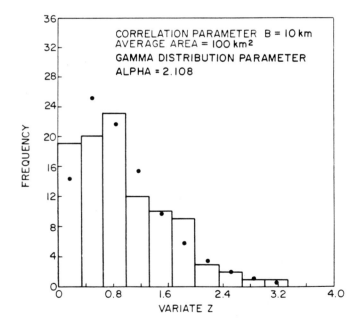

Figure 4a: Areal averaging of weakly correlated intensity over a small catchment.

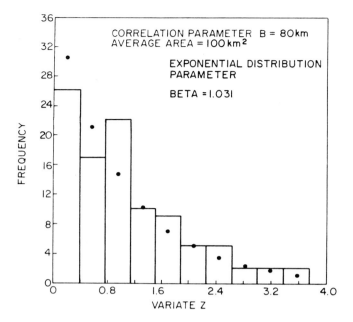

Figure 4b: Areal averaging of strongly correlated intensity over a catchment.

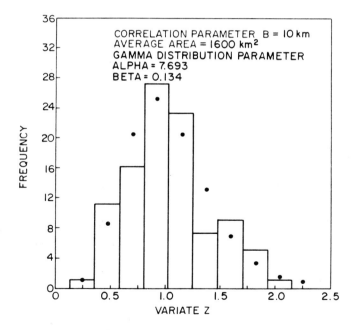

Figure 5a: Areal averaging of weakly correlated intensity over a large catchment.

TABLE 2: Additional Parameters for Areal Rainfall Effects

Common Parameters

PARAMETER	SPECIFICATION
velocity v (m/s)	1
duration scale λ (hr)	7.5
point intensity scale α'	1
point intensity shape $\beta^{*\prime}$	1
runoff ratio r	1

Catchment Parameters

PARAMETER	SPECIFICATION	
	SCa	LCa
order Ω	3	5
area $A_T\,(km^{2})$	100	1600
length $L_\Omega(km)$	10.31	23.19
duration scale λ^{*}	2.63	1.15

Areal Climatic Parameters

PARAMETER	SPECIFICATION			
	SCl		LCl	
correlation scale b (km)	10		80	
	SCa	LCa	SCa	LCa
correlation scale b^{*}	1.0	.125	8.0	2.0
intensity scale β^{*}	.453	.134	1.031	0.981
intensity shape α	2.108	7.693	1.0	1.0

SCa - small catchment   LCa - large catchment
SCl - small climate   LCl - large climate

from the point process while for small b , the areal process should be more like the limiting gamma. It should be noted that the limiting gamma distribution is determined by the number of points n and the point distribution parameters. For large n and $b = 0$, the areal process tends to the normal distribution, a consequence of the Central Limit Theorem.

The complete specification of the derived distribution still requires values for storm duration scale λ and velocity v , for which values of $\lambda = 7.5$ hours and $v = 1$ m/s were used. The λ value is characteristic of Boston in the northeast U.S. while the velocity is a reasonable value for flood flows. All of the additional data for the areal averaging example is summarized in Table 2. The results of averaging small and large climates over the small catchment are presented in Figures 4a and 4b, respectively. Figure 4a shows the sample and fitted distributions for the small climate. The effects of averaging are obvious, with the areal distribution assuming a marked humped appearance ($\alpha = 2.108$, $\beta^{*} = 0.453$). In contrast, the large climate/small catchment areal distribution in Figure 4b ($\beta^{*} = 1.031$) is essentially identical

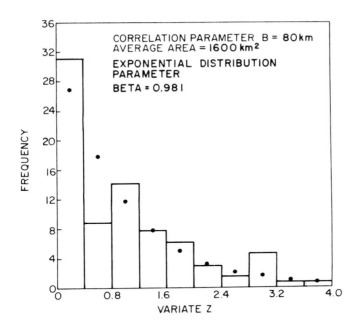

Figure 5b: Areal averaging of highly correlated intensity over a large catchment.

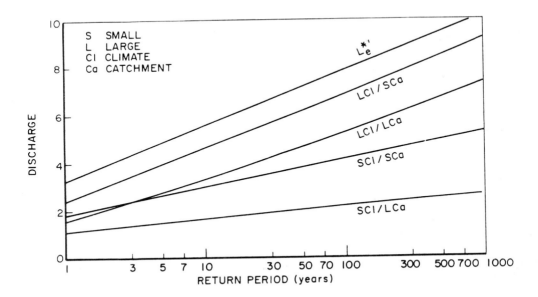

Figure 6:   Effects of rainfall averaging on derived frequency curves.

to the underlying point process. The results for large and small climates over the large catch-ment in Figures 5a and 5b are similar to the small catchment results. The large climate/large catchment area distribution is exponential ($\beta^* = .981$) but the averaging effects for the small climate are even more pronounced over the large catchment than over the small catchment ($\alpha = 7.693$, $\beta^* = .134$). Even though the dimensionless frequency was derived for $\beta^* = 1$ the effects of averaging must still be retained in the areal scale parameter as well as the shape parameter α. That is why both the simulation sample scale and shape parameters are reported here. It makes little sense to compare the results from two different catchments and climates on the basis of a single $\beta^* = 1$. At the same time it is important that only *relative* scale effects be considered. In the current example a dimensionless standard exponential point process ($\beta' = 1$) was used so that the dimensionless areal intensity scale parameters are just the sample values from the simulation. In the case of real data, physical area scale parameters should be normalized by a common standard scale value so that some feeling for relative scale differences is retained. The derived moment ratios are not affected by β^* but the mean and position of the frequency curve are.

The implications of areal averaging are significant since the input areal intensity distribution constitutes one of the two limiting forms of the derived frequency. This limit is approached at high λ^* values (long storms relative to response time). From Figure 2 it appears that for smaller λ^* (larger catchments if all other parameters are held constant) the derived frequency skew coefficient C_s mildly increases. This would appear to contradict real-world experience where small catchments often display higher C_s than larger catchments. However, Figure 2 should not be used to infer effects of catchment scale since the effects of catchment size on areal averaging and the intensity distribution parameters were not included. Averaging causes the skew of the areal distribution to be less than the skew of the underlying point pro-cess, thus reducing the limiting C_s attainable in the derived frequency. This reduction in C_s is effective over the whole range of λ^* so that the mild increase in C_s due to decreasing λ^* is counteracted by a decrease in C_s due to averaging. The averaging effect should dominate if the point process is moderately or weakly correlated relative to the catchment scale. This suggests that some of the observed differences in flood frequency skewness between small and large catchments is attributable to the interaction of catchment and climatic spatial scale.

The frequency curves for the four catchment/climate combinations are plotted in Figure 6 along with the underlying point intensity frequency curve. The derived moments are sum-marized in Table 3. Consider first the curves for the large (LCl) and small (SCl) climates on the small catchment (SCa). The difference between the limiting curve and the SCl/SCa curve is due solely to the rainfall-runoff transformation since the areal averaging effects are negligible. The slopes of the point intensity and SCl/SCa frequency curves are the same but the limit curve lies higher. The LCl/SCa curve has a flatter slope, indicating a smaller C_s than for the SCl/SCa curve. The differences between the LCl/LCa curve and the point intensity frequency are due to both areal averaging and the rainfall-runoff transformation. A similar pattern is observed for the small and large climates on the large catchment (LCa). However, for a given climate the large catchment curve lies below the small catchment curve. This comparison between curves for a given climate is informative and reassuring. It shows that when rainfall areal averaging is considered the smaller catchments tend to be flashier and more highly skewed than their larger counterparts, all else held constant. The fact that the derived frequency curves all lie below the limiting intensity frequency curve reflects on the damping action of the rainfall-runoff transformation.

The areal averaging examples were calculated on the basis of site-specific physical parameters. However, the problem can be posed in a dimensionless manner consistent with the dimension-less derived frequency if the correlation parameter b is normalized by a suitable characteristic

TABLE 3. Frequency Moments for Areal Averaging Example

case	μ	C_V	C_S
point intensity	4.22	0.24	2.00
SCl/SCa	2.31	0.22	1.93
LCl/SCa	3.35	0.29	2.08
SCl/LCa	1.37	0.17	1.69
LCl/LCa	2.31	0.34	2.26

SCl - small climate	LCl - large climate
SCa - small catchment	LCa - large catchment

length. Rodriguéz-Iturbe and Mejia (1974a) showed that on geometrically similar areas the expected correlations between two randomly chosen points in the areas A_1 and A_2 with correlation parameters b_1 and b_2, respectively, are identical if the quantities (A_1/b_1^2) and (A_2/b_2^2) are equal. Alternatively, this similarity condition can be phrased in terms of a dimensionless correlation parameter b^*. Given a characteristic length a of the area, if $b_1^* = b_1/a_1$ and $b_2^* = b_2/a_2$ are identical then correlation similarity is preserved. The choice of characteristic length is somewhat arbitrary and may be dictated by the shape under consideration -- it might be a side length, diagonal, diameter, etc. In any event the characteristic length will be of the form $cA^{.5}$, where c is a constant of proportionality dependent only on the shape. In the dimensionless context, then, the transformation from point rainfall (unit scale parameter $\beta^{*'}$ assumed) to areal rainfall depends on the dimensionless correlation parameter b^*. The parameters (α, β^*) of the resultant areal distribution depend on both the point parameters and b^*. The dimensionless frequency can thus be characterized by the parameter set $(\alpha', b^*, \lambda^*, \kappa, R_L, n)$.

5. CONCLUSIONS

A dimensionless flood frequency has been derived from a direct runoff IUH and independent distributions of rainfall intensity and duration. By relating the IUH model to observable catchment characteristics and accounting for the spatial correlation of the rainfall process it is possible to determine the roles of catchment and climatic scale in flood frequency response. Catchment and climatic scales interact in two ways: 1) through the relative magnitudes of storm duration and catchment response time as captured in the dimensionless parameter λ^*; 2) through the extent of rainfall spatial correlation relative to catchment size as embodied in the dimensionless parameter b^*. The λ_* parameter represents the effect of the rainfall-runoff transformation on the input areal intensity distribution., Large λ^* indicates very little change in the output flow distribution as compared the intensity input; small λ^* indicates the opposite. The b^* parameter represents the effect of areal averaging on the underlying point intensity distribution. Large b^* indicates a minor difference between the point and averaged processes; significant differences are indicated by small b^*. When λ^* is taken as a function of catchment scale only, catchment scale imposes two opposing tendencies on flood frequency behavior. Through λ^*, increasing scale produces higher skew coefficient; through the averaging process increasing scale reduces the skewness of the limiting area intensity distribution. This latter effect appears to be dominant and thus the implications of the derived frequency are consistent with real-world data, since flood frequency behavior of small catchments tends to be both flashier and more highly skewed than the behavior of large catchments.

These findings suggest that the parameters λ^* and b^* could be used as criteria in consideration of flood frequency similarity. The derived distribution approach could also be used to explore the effects of seasonal climate in flood frequency response. In this study a single homogeneous climate was assumed. However, it is a simple matter to calculate the frequencies for several climates and then combine them according to the rules of mixed distributions (Haan, 1977). The resultant effects could be considerable in view of the areal averaging results for highly correlated (frontal storms, typical of the eastern U.S. autumn months) and weakly correlated (convective storms, typical of summer months) point rainfall processes.

ACKNOWLEDGMENTS

This work was suported in part by the National Science Foundation Grant CME-79-15168, called "The Development of a Stochastic-Conceptual Hydrologic Model for Analyzing the Statistical Response of Flow Generation and Flood Dynamics". Dr. Hebson was also supported by the State of New Jersey Garden State Fellowship Program during the execution of this work.

REFERENCES

Boyd, M.J., 1978. "A Storage-Routing Model Relating Drainage Basin Hydrology and Geomorphology", *Water Resources Research*, 14(5), pp. 921-928.

Chan, S. and R.L. Bras, 1979. "Urban Stormwater Management: Distribution of Flood Volumes", *Water Resources Research*, 15(2), pp. 371-382.

Cordova, J.R. and R.L. Bras, 1981. "Physically Based Probabilistic Models of Infiltration, Soil Moisture, and Actual Evapotranspiration", *Water Resources Research*, 17(1), pp. 93-106.

Corey, A.T., 1986. *Mechanics of Heterogeneous Fluids in Porous Media*, manuscript to be published.

Dias-Granados, M.A., J.B. Valdés, R.L. Bras, 1984. "A Physically Based Flood Frequency Distribution", *Water Resources Research*, 20(7), pp. 995-1002.

Eagleson, P.S., 1978. "The Distribution of Annual Precipitation Derived from Observed Storm Sequences", *Water Resources Research*, 14(5), pp. 713-721.

Eagleson, P.S., 1972. "Dynamics of Flood Frequency", *Water Resources Research*, 8(4), pp. 878-898.

Eagleson, P.S., 1970. *Dynamic Hydrology*, McGraw-Hill, NY.

Freeze, R.A., 1980. "A Stochastic-Conceptual Analysis of Rainfall-Runoff Processes on a Hillslope", *Water Resources Research*, 16(2), pp. 391-408.

Gray, D.M. and J.M. Wigham, 1973. "Peak Flow - Rainfall Events" in *Handbook on the Principles of Hydrology* (Gray, ed.), Water Information Center, Huntington, NY.

Haan, C.T., 1977. *Statistical Methods in Hydrology*, Iowa State University, Ames, Iowa.

Hebson, C.S. and E.F. Wood, 1982. "A Derived Flood Frequency Distribution Using Horton Order Ratios", *Water Resources Research*, 18(5), pp. 1509-1518.

Henderson, F.M., 1963. "Some Properties of the Unit Hydrograph", *Journal of Geophysical Research*, 68(16), pp. 4785-4793.

Kibler, D.F., 1982. "Desk Top Methods for Urban Storm-Water Calculations", in *Urban Stormwater Hydrology* (Kibler, ed.), American Geophysical Union, pp. 87-136, Washington, DC.

Klemes, V., 1978. "Physically Based Stochastic Hydrologic Analysis", in *Advances in Hydroscience* (Chow, ed.), Vol. 11, pp. 285-356, Academic Press, NY.

Leclerc, G. and J.C. Schaake, 1972. "Derivation of Hydrologic Frequency Curves", *Tech. Report No. 142*, Ralph Parsons Laboratory for Water Resources and Hydrodynamics, MIT, Cambridge, MA.

Lindgren, B.W., 1976. *Statistical Theory*, Macmillan, NY.

Loganathan, G.V. and J.W. Delleur, 1984. "Effects of Urbanization on Frequencies," *Water Resources Research*, 20(7), pp. 857-865.

Mantoglou, A. and J.L. Wilson, 1982. "The Turning Bands Method for Simulation of Random Fields Using Line Generation by a Spectral Method", *Water Resources Research*, 18(5), pp. 1379-1394.

Nash, J.E., 1957. "The Form of the Instantaneous Unit Hydrograph", *IAHS/AISH* Pub. 42, pp. 114-118.

Nash, J.E., 1960. "A Unit-Hydrograph Study with Particular Reference to British Catchments", *Proceedings of the Institute of Civil Engineers*, 17, pp. 249-283.

Ott, R.F., 1971. "Streamflow Frequency Analysis Using Stochastically Generated Hourly Rainfall", *Tech. Report No. 151*, Dept. of Civil Engineering, Stanford University, Palo Alto, CA.

Philip, J.R., 1957. "The Theory of Infiltration", 1-7, *Soil Sciences* Volumes 83, 84, 85.

Rodríguez-Iturbe, I., M. Gonzalez Sanabria, and R.L. Bras, 1982a. "A Geomorphoclimate Theory of the Instantaneous Unit Hydrograph", *Water Resources Research*, 18(4), pp. 877-886.

Rodríguez-Iturbe, I., M. Gonzalez Sanabria, and G. Caamaño, 1982b. "On the Climatic Dependence of the IUH: A Rainfall-Runoff Analysis of the Nash Model and the Geomorphologic Theory", *Water Resources Research*, 18(4), pp. 887-903.

Rodríguez-Iturbe, I. and J. Valdés, 1979a. "The Geomorphological Structure of Hydrologic Response", *Water Resources Research*, 15(5), pp. 1409-1420.

Rodríguez-Iturbe, I., G. Devoto and J. Valdés, 1979b. "Discharge Response Between the Geomorphologic IUH and the Storm Characteristics", *Water Resources Research*, 15(6), pp. 1435-1444.

Rodríguez-Iturbe, I. and J.M. Mejia, 1974a. "The Design of Rainfall Networks in Time and Space", *Water Resources Research*, 10(4), pp. 713-728.

Rodríguez-Iturbe, I. and J.M. Mejia, 1974b. "On the Transformation of Point Rainfall to Areal Rainfall", *Water Resources Research*, 10(4), pp. 729-735.

Rosso, R., 1984. "Nash Model Relation to Horton Order Ratios", *Water Resources Research*, 20(7), pp. 914-920.

Smart, J.S., 1972. "Channel Networks" in *Advances in Hydroscience* (Chow, ed.), 8, pp. 305-346, Academic Press, NY.

Stol, P.H., 1972. "The Relative Efficiency of Rain-Gauge Networks", *Journal of Hydrology*, 15, pp. 193-208.

Surkan, A.J., 1969. "Simulation of Storm Velocity Effects on Flow from Distribution Channel Networks", *Water Resources Research*, 5(1), pp. 112-128.

U.S. Weather Bureau, 1958. "Rainfall Intensity-Frequency region 2, Sotheastern United States", *Tech. Paper 29* U.S. Dept. of Comm., Washington, DC.

Wood, E.F., 1976. "An Analysis of the Effects of Parameter Uncertainty in Deterministic Hydrologic Models", *Water Resources Research*, 12(5), pp. 925-935.

Wood, E.F. and B.M. Harley, 1975. "The Application of Hydrologic Models in Analyzing the Impact of Urbanization", Proceedings of the Bratislava Symposium, *IAHS Pub. No. 115*.

Wood, E.F. and C. Hebson, 1986 (in press). "On Hydrologic Similarity: 1. Derivation of the Dimensionless Flood Frequency Curve", *Water Resources Research*, (22).

8

ON SCALES, GRAVITY AND NETWORK STRUCTURE IN BASIN RUNOFF

Vijay K. Gupta
Ed Waymire
Ignacio Rodríguez-Iturbe

ABSTRACT

Runoff generation and its transmission to the outlet from an ungaged river basin having an identifiable channel network are considered at the basin scale. This scale is much larger than the hydrodynamic scale, where the equations governing the transport of water overland and in saturated and unsaturated soils are best understood. Gravity, via altitude, plays the fundamental role in both the transport of water as well as in network formation via erosion and sediment transport. So, here altitude is identified as the natural parameter for physically rigorous descriptions of network structures in the context of hydrologic investigations at the basin scale. In this connection an empirical postulate is made on the link heights as being independent but possibly non-homogeneous random variables having an exponential distribution. Data from six river basins ranging in sizes from 1 sq. km to 100 sq. km and from different climatic regions are used to test the suitability of this postulate. The drainage scaling parameter D_N is introduced as the number of links per unit area density in an infinitesimal increment of the altitude at the basin scale. Data from five of the six basins is analyzed to show qualitatively that these basins are homogeneous with respect to D_N. This homogeneity along with that in the exponential nature of the link heights are used to illustrate that the total runoff generated by the sub-basin associated with any link of a basin, has a gamma distribution with parameters $\lambda/\bar{\mu}$ and 2(link magnitude) - 1 . ,The parameters denoting the link magnitude, the mean link height λ^{-1}, and the long time average volume of runoff per unit elevation of a link, $\bar{\mu}$, are meaningful only at the basin scale.

1. INTRODUCTION

Quantitative understanding and prediction of the processes of runoff generation and its transmission to the outlet, i.e., the hydrologic response, in ungaged basins represent two basic

V. K. Gupta et al. (eds.), Scale Problems in Hydrology, 159–184.
© *1986 by D. Reidel Publishing Company.*

problem areas in scientific hydrology. The most serious handicap to progress in these areas is the simple fact that the best developed hydrology theories, generally applicable at small space-time scales, cannot simply be extrapolated to quantify hydrologic processes at the large basin wide scales at which predictions are generally sought.

Current hydrologic theories describing the movement of water, be it overland, in channels, in vadose zones, or in aquifers, have been developed predominantly at the **hydrodynamical scale** in homogeneous materials; largely from experimental studies in parking lots, flumes and laboratory columns. However, in a river basin spatial heterogeneities are present everywhere in terms of soil characteristics, topography, vegetation cover, etc., both within hillslopes and between the hillslopes along the channels of a network. Field observations within the last decade have shown that these spatial heterogeneities produce incredibly complex patterns of runoff generation on hillslopes (Hewlett, 1961; Dunne, 1982; Pilgrim, 1983). Since there is no known empirical evidence to suggest that all the physical details at the hydrodynamical scale are also dominant in the collective hydrologic response from a hillslope, as an alternative to deductions based on differential equations formulations at the hydrodynamic scale, in recent years this view has led to attempts to directly quantify runoff generation over a hillslope via simple phenomenological equations (Beven, 1986; Kirkby, 1986). Of course, possible theoretical deductions of these formulations from differential equations at the hydrodynamic scale remain a distant goal. Such formulations governing the collective hydrologic response from a hillslope can be viewed as applicable over a **hillslope scale**. As one extends the scale of inquiry from a single hillslope to that of a basin, possessing an identifiable channel network structure, one immediately encounters several hillslopes governing runoff generation along the' links of the network. A priori one expects that the individual hillslopes within the basin vary in their physical characteristics. Consequently, an equation governing runoff generation over any one hillslope need not be applicable to the others in the system. Yet in the context of the collective generation of runoff from a basin and its hydrograph at the basin outlet, one can speculate that all the variabilities among individual hillslopes are not dominant. Therefore, in order to quantify the hydrologic response of a basin possessing an identifiable channel network structure, it seems natural to consider an area within a basin, which encloses a small portion of the basin network. The runoff generation from such an area will be different from that of a single hillslope. Partitioning a basin into such areas for quantifying basin response leads us to define the notion of a **basin scale**, as discussed in section 4.

The field observations of rainfall-runoff over small plots of hillslope regions exhibit tremendous complexities and spatial variabilities at the hydrodynamical scale, (see Pilgrim, 1983) and seem to defy any physically general theoretical formulation. However, the collective description of runoff generation from the system of hillslopes and its transmission via the channel network to the basin outlet may admit surprisingly simple descriptions if viewed on the basin scale. Our idea behind understanding these descriptions in a physically general and mathematically rigorous manner is not to start with formulations either at the hydrodynamical scale or at the hillslope scale, but rather it is to search for physical "laws" governing these descriptions directly at the scale of interest, i.e., the basin scale. This is not meant to imply that lower scale problems are of no interest. To the contrary, a good measure of the depth of scientific understanding of natural phenomena is the number of different scale ranges over which the phenomena can be consistently deduced at one scale from the one below it.

In section 2 we illustrate explicitly how the network geometry contributes to the structure of the network instantaneous unit hydrograph (IUH). A fundamental role in this description is played by the channel link structure and in particular the width function $N(x)$, representing the number of links at flow distance x, discussed in Kirkby (1976). Moreover, the link structure also connects a network with the hillslope. Various properties of the width function have been studied in detail in Mesa (1982) and in Troutman and Karlinger (1984, 1985, 1986). A flow equation governing the IUH from a single network is also derived which explicitly reflects the role of "averaging" with respect to the network structure as well as transport dynamics from various parts of the network to the outlet. Applications of special cases of the equation to an understanding of basin IUH are discussed further in Kirkby (1976) and Mesa and Mifflin (1986).

In spite of the steady progress made towards the goal of understanding hydrologic response at the basin scale alluded to above, the issue regarding the manner in which the network geomorphology enters into our understanding of basin response has remained unsettled on fundamental physical grounds, particularly because the role of gravity via altitudes in network description has thus far not been clearly identified. Gravity most certainly is fundamentally important with regard to runoff generation over hillslopes, the development of channels via erosion and sediment transport, and the transport of water through the network. In spite of this, heretofore the altitudes could only be incorporated in response characteristics through a back door approach, e.g., in the phenomenological descriptions of runoff generation on a single hillslope or via a slope parameter in the routing flow equation for the network (Rodriguéz-Iturbe et al., 1982; Mesa and Mifflin, 1986). Since a river basin can be viewed as consisting of two interrelated physical systems, the hillslopes and the channel network, the uncovering of reciprocal controls between these two systems in the form of laws governing the structural regularity between them lies at the heart of understanding and quantifying the hydrologic response at the basin scale. These ideas are briefly discussed in section 3.

The remaining parts of the paper, comprising sections 4 and 5, represent a first step in the exploration of possible unifying principles for the description of channel network characteristics involving gravity via altitudes. Indeed, as is discussed in section 4, altitudes play the fundamental role of a "parameter" in the network description; just as "time" typically appears as a parameter in various descriptions of physical systems. An empirical regularity with regard to the "link heights or drops" as being independent and possibly non-homogeneous random variables with an exponential density function is postulated and tested with the help of data from six basins from different climates, geology, etc. The notion of basin scale is explicitly discussed in this section. Section 5 explores the analytical structure of the process $N(h)$, denoting the number of links at the altitude h, and henceforth called the "link concentration function". An analytical result is given in an attempt to identify structural regularity in the fluctuation of the runoff volume generated by the hillslopes along a network as dictated by the process $N(h)$. In particular, it is shown explicitly how the parameters representing network magnitude, mean link height, etc., arise quite naturally in the large scale description of fluctuations in runoff generation in a basin. The paper is concluded with brief remarks on future research directions in this important area of hydrology.

2. CHANNEL LINK STRUCTURE AND THE GEOMORPHOLOGIC IUH

Our objective in this section is to derive an equation for the network response which employs the notion of "averaging" in the equation governing water transport from any point in the network to the outlet, as well as in the bifurcating tree structure of the channel network. Since this flow equation is meaningful for a basin having a well defined channel network structure, it represents a point of departure from a flow equation in a single open channel.

Consider the usual bifurcating tree representation of a network as introduced by Shreve (1966). A fundamental property of this network in the plane is that there is a unique one dimensional path connecting any pair of points in the tree. In particular the flow path from any point to the basin outlet is uniquely determined as a one dimensional path, as shown schematically in Fig. 1. It is interesting that even here there is an implicit use of altitudes in identifying the outlet among all the external links. Moreover, the notion of network order, itself, can easily be seen to depend on how this selection is made; i.e., order implicitly depends on "altitude" even in a planar representation of a channel network.

In the ensuing analysis it is only the basic tree geometry which is exploited. There is no appeal to the Horton's laws or Shreve's postulates (Shreve, 1966) with regard to the network structure. Let $g(t,x)$ denote the response at the outlet at time t due to a delta function input applied at instant 0 at a flow distance x from the basin outlet. One may interpret $g(\cdot,\cdot)$ as a space-varying partial instantaneous unit hydrograph (IUH); space-varying because it depends on x and partial because it is not the IUH of the entire network but rather the response function of a small local segment of the network. The product $g(t,x)dt$ then represents the probability that a particle injected in the network at time 0 at a point at flow distance x from the outlet will reach the outlet in the time interval $[t,t+dt]$. We now assume that all points at equal flow distances x from the outlet have the same response function $g(t,x)$. This assumption reflects an "averaged" flow equation for the network in the following sense. Whereas various channels at a fixed distance x from the outlet would have different cross sectional areas and other geometric and roughness characteristics, the above assumption implies that they cannot be distinguished from one another with respect to their response at the outlet. Although a precise determination of this function $g(t,x)$ is an illusive problem, for the present we assume that $g(t,x)$ satisfies the following convolution (semi-

$$g(t,x+y) = \int_0^t g(t-s,x)\, g(s,y)\,ds, \quad x,y \geq 0 \tag{1}$$

The semi-group condition is natural in the context of the interpretation of an IUH as the probability density of the holding time given in Rodriguez-Iturbe and Valdes (1979), Gupta et al., (1980) and Gupta and Waymire (1983). Three examples of kernels which satisfy the semi-group property are:

$$g(t,x) = \delta(t - x/v) \tag{2}$$

$$g(t,x) = \alpha(\alpha t)^{x-1}\, e^{-\alpha t} / \Gamma(x), \quad t \geq 0 \tag{3}$$

$$g(t,x) = \frac{x}{(4\pi D t^3)^{1/2}} \exp\left\{\frac{(x-vt)^2}{4Dt}\right\}, \quad t \geq 0 \tag{4}$$

In Eqns. (2), (3), and (4), the parameters, v, α and D are positive scalars.

Various geomorphologic quantities required for the present development are (i) $N(x)$, the width function denoting the number of links at a flow distance x from the outlet, and (ii) Z, the sum of all link lengths in the network. It is important here to distinguish the notion of geometric link length from that of the flow link length. The former definition used by Shreve (1969) denotes the straight line distance between two junctions or a junction and a source defining a link, whereas the latter denotes the length along the channel between two junctions of a link or a junction and a source. The schematic diagram in Fig. 1 depicts these two definitions of link length. First note that if we take the link length to be the geometric link length, then Horton's law of stream numbers and stream lengths yield the following approximate expression for Z (see, e.g., Eagleson, 1970, p. 386):

$$Z = L_\Omega\, [(R_B/R_L)^\Omega - 1] / [(R_B/R_L) - 1] \tag{5}$$

where R_B is Horton's bifurcation ratio, R_L the stream length ratio, Ω the basin order and L_Ω the length of the highest order channel. The following developments are valid for either of

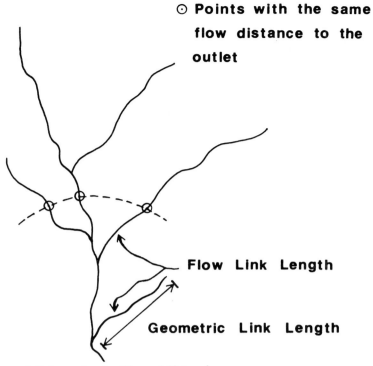

⊙ **Points with the same**
flow distance to the
outlet

Flow Link Length

Geometric Link Length

Figure 1   A Schematic of a Channel Network

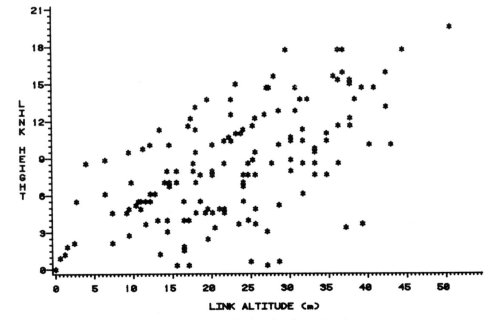

Figure 2   Heights versus Altitudes of Links for Goodwin Creek
Sub-basin 2.

these two definitions of a link length unless explicitly stated otherwise.

To obtain a representation of the geomorphologic IUH for the network, assume that a unit depth of runoff is generated instantaneously and uniformly over the network. Then the probability that a randomly, i.e., uniformly over the network, injected drop will fall between a flow distance x to $x + dx$ is given by $Z^{-1}N(x)dx$. The probability that this drop will reach the outlet between t to $t + dt$ is then $Z^{-1}N(x)g(t,x)dx$. Summing over the x's leads to a representation of $Q(t)$, the network IUH,

$$Q(t) = \int_0^\infty Z^{-1} N(x) \, g(t,x) \, dx. \tag{6}$$

Note that $Q(t)dt$ represents the probability that a randomly injected drop over the network arrives at the outlet in the interval t to $t + dt$.

Several particular cases of Eqn.(6) are of special interest. As an example, consider the case when all particles travel at the same speed v, assumed to be constant throughout space and, for convenience here, also time, on their way to the basin outlet. This is the case given by Eqn. (2). According to Eqn. (2) a particle reaches the outlet at time $t = x/v$ with probability one. In this particular example the network IUH is then given by:

$$Q(t) = \int_0^\infty \delta(t - x/v) \, Z^{-1} \, N(x) \, dx$$

$$= \int_0^\infty v \, \delta(t - y) \, Z^{-1} N(yv) \, dy$$

or

$$Q(t) = Z^{-1} \, v \, N(vt) \tag{7}$$

This important special case is discussed in Kirkby (1976). As illustrated in Kirkby (1976) it has immediate practical application for any given basin and is extremely simple. There can be absolutely no misunderstanding of the system being described by Eqn. (7). The geomorphological parameters involved can easily be read off a map and more importantly, it illustrates an explicit appearance of the network geometry in the expression for the IUH. The connection of Eqn. (7) with Horton's laws can be accomplished through Eqn. (5) if we apply the geometric definition of link lengths. The function $N(\cdot)$ can be computed from the topographic map of a basin (see, Kirkby, 1976; Mesa and Mifflin, 1986) and Eqn. (7) can then be applied under different velocity scenarios.

The general formulation represented by eq. 6 also allows for the study of the effects of other types of choices for the function $g(t,x)$ on the network IUH. The study by Mesa and Mifflin (1986) gives an application of Eqn. (6) to a real world basin and provides physical insights into the relative role of the channel network IUH and the hillslope response on a basin IUH.

An important problem is to determine the way in which the bifurcation structure of a network is reflected in the basin IUH defined by Eqn. (6). To illustrate the nature of this problem, let us call x_0 the root length or length of the link from the outlet to the first junction. Calling $N^{(1)}(x - x_0)$ and $N^{(2)}(x - x_0)$ the width functions of the left hand and the right hand side networks when the origin is taken to be x_0 at the top of the root link, one may write Eqn. (6) as,

$$Q(t) = \int_0^{x_0} g(t,x) \, Z^{-1} \, N(x) \, dx + \int_{x_0}^{\infty} g(t,x) \, Z^{-1} \, N(x) \, dx$$

$$= Z^{-1} \int_0^{x_0} g(t,x) N(x) \, dx$$

$$+ Z^{-1} \int_{x_0}^{\infty} g(t,x) \, [N^{(1)}(x - x_0) + N^{(2)}(x - x_0)] \, dx \qquad (8)$$

Making the change of variables $y = x - x_0$ in Eqn. (8) and applying the semi-group property given by Eqn. (1), we observe that the second term on the right hand side of Eqn. (8) consists of two double integrals, one of which is identical to Eqn. (6). If $Z^{(1)}$ and $Z^{(2)}$ denote the total channel lengths of the left hand and the right hand side networks, then Eqn. (8) can be written as

$$ZQ(t) = \int_0^{x_0} g(t,x) \, dx$$

$$+ \int_0^{t} [Z^{(1)} \, Q^{(1)}(s) + Z^{(2)} \, Q^{(2)}(s)] g(t - s, x_0) \, ds \qquad (9)$$

Here $Q^{(i)}, i = 1,2$, are the left and right IUH of the subnetworks at the first junction from the outlet. One could now disaggregate $Q^{(1)}(t)$ and $Q^{(2)}(t)$ into two different components in exactly the same way as was done for $Q(t)$. The disaggregation could then continue to cover the whole network. From the ensemble point of view, the derived flow Eqn. (9) may be regarded through its continuous disaggregation as a generalized renewal equation imposed by the network on the distribution of the stochastic flow process $Q(t)$. In other words, from the ensemble point of view it is the structure of the probability distribution of the IUH which is of interest. The basic probabilistic assumption for the stochastic process $ZQ(t)$ would be that $ZQ = Z^{(0)} Q^{(0)}, Z^{(1)} Q^{(1)}, Z^{(2)} Q^{(2)}, \ldots$ are statistically similar; i.e., removal of an external link does not effect the statistical structure of ZQ in this ensemble.

One application of the basin ensemble view outlined above is the investigation of issues concerning similarity in the flow $\{Q(t)\}$ given various large scale basin parameters such as order Ω, magnitude M, level L, etc. From the point of view of least squared error prediction, the best predictor is the conditional expectation of $Q(t)$ given the values of these large scale parameters. Expectation and other moments of $(Q(t))$ for the basin ensemble can be investigated via Eqn. (9). Troutman and Karlinger (1985) consider the problem of calculating best predictors of a flow given large magnitude M and/or high level L basins. However, the network flow equation studied in Troutman and Karlinger (1985) is not the same as given by Eqn. (9) and it is not clear that there is even a connection. Nevertheless, in view of the hydrologic interpretation of the width function $N(\cdot)$ it is clear that the results in Troutman and Karlinger (1984) for the width function do apply at least to the special case given by Eqn. (2). In this connection there is an interesting unresolved problem which can be understood as follows.

There are two extreme cases for the basin evolution represented by: (i) bifurcation at every interior vertex, and (ii) one and only one bifurcation at each interior vertex. In case (i) one observes an exponential relationship between M and L of the form

$$M = 2^L \tag{10}$$

On the other hand in case (ii) the relationship is linear and of the form

$$M = L - 1 \tag{11}$$

It is the latter case which has been solved by Troutman and Karlinger (1984). However, if one considers the fact that for the most likely basins the main channel length and the area exhibit a power law relation (Shreve, 1974), then one might expect a relationship of the form

$$M = L^\theta + C \tag{12}$$

as a natural case to consider. The effect of conditioning on such basin structure is to assign zero weight to basins of an a-typical character. Analytic results for Eqn. (12) along similar lines as given in Troutman and Karlinger (1984) may be quite different if $\theta \neq 1$.

Apart from obtaining the average characteristics of the flow, some idea with regard to the fluctuations will be required for the averages to be meaningful; see for example Gupta and Waymire (1983, Theorem 1, p. 116). In this connection both the mean and the variance for the center of mass of the width function in the case of constant link lengths have been calculated by Mesa (1982). Some interesting results on fluctuations around the mean IUH are given in Troutman and Karlinger (1986).

3. ALTITUDE: THE MISSING DIMENSION IN LINKING NETWORKS WITH BASIN HYDROLOGY

The developments of the last section dealt with a network IUH rather than a basin IUH because the instantaneous unit runoff input was assumed to be distributed uniformly over the network, rather than over the whole basin. This in physical terms is tantamount to the following assumptions: (1) Direct runoff flow time from the hillslopes to the network is zero, (2) The amount of the stormflow contribution from any hillslope is measured by the length of the channel link at the base of the hillslope. The first assumption may be reasonable as long as one is dealing only with the overland flow component of the direct runoff. However, the subsurface flow component of the stormflow has a delay time to reach the network which cannot be neglected (see, e.g., Mesa and Mifflin, 1986). The question, then, concerns the importance of the subsurface component. This question is linked with the second assumption which involves the amount of runoff produced by each hillslope. No adequate answer is yet available for this question. Our view towards solving this problem is to look for structural regularity among the runoff producing areas of the hillslope system along the channel network via structural regularity in the network. This represents a point of departure from the traditional attempts at rainfall-runoff modeling of each hillslope separately using hydrodynamical equations, or modeling only one hillslope phenomenologically and then extrapolating the results to others on an ad hoc basis. Our idea is motivated by the various hydrologic considerations discussed below.

For the purpose of this discussion we consider the basin as made up of two interrelated systems: the channel network and the hillslopes. The hillslopes control the production of storm water runoff which, in turn, is transported through the channel network towards the basin outlet. The runoff contributing areas of the hillslopes are both a cause and an effect of the drainage network growth and development. This cause and effect relationship may be visualized through the following considerations. In short and intense rainfall storms most of the runoff is generally contributed by either Hortonian overland flow or by direct precipitation into the saturated areas around the channels; which of these two mechanisms prevails in a given situation depends on the soil and vegetation characteristics regulating the infiltration capacity. The second mechanism produces what is called the saturated overland flow and was first articulated cogently by J.D. Hewlett and A.R. Hibbert (Hewlett, 1961; Hewlett and Hibbert, 1967). When rainfall is not very intense but has a long duration, contributions of subsurface storm flow occur mainly from areas close to the drainage network. In such long storms the contributions of saturated overland flow are also very important and may increase more rapidly than those of the subsurface stormflow. The saturated areas expand rapidly from poorly drained soils into areas of initially better drained soils and steeper topography. As explained by Dunne (1978), hollows are preferred avenues for such expansions, partly because of their role in concentrating downslope surface and subsurface flow, but to a much greater extent, because their concave profiles cause the water table to be closer to the surface at the beginning of this storm. As pointed out by Rodriguez-Iturbe (1979) the convergence of overland flow is a major cause of the growth upslope of the channels making up the network, but on the other hand, hollows are a cause of the occurrence of overland flow. From this point of view the drainage network itself may be seen as a reflection of the runoff producing mechanisms occurring in a basin. The discovery of structural regularity among the runoff producing areas of the basin hillslopes is a crucial and fascinating hydrologic problem. Indeed there has been a continuous advancement in the understanding of the physical processes which control the response of a given hillslope to a precipitation input under certain simplifying assumptions of homogeneity in soil and vegetation characteristics (Kirkby, 1978). Nevertheless basins are made up of a very large number of hillslopes, and the hillslopes can display a large overall variability of soil and vegetation properties. The knowledge of the principles which control the behavior of a single hillslope-which is both necessary and valuable-by itself cannot be expected to lead to an understanding of the role of the hillslope system in the hydrologic response of a basin.

From the previous discussion the existence of circuits of reciprocal control between the system of hillslopes and the drainage network of a basin is apparent. One expects a whole interlocking system, as yet scarcely studied, which commands growth and differentiation in the drainage network. The importance of these reciprocal controls lies at the very heart of hydrology. Every branch of the network is linked to a downstream branch for the transportation of water and sediment but it is also linked for its viability-through the hillslope system-to another branch and in a sense, to every other branch. This second linkage through the hillslope system results from the fact that in order for a stream channel to be maintained, the hillslope area generating runoff must not fall below a certain minimum.

In the above interlocking system the drainage network should then be seen as the pattern which connects, in the sense that it relates the different parts of the basin between themselves and also in combination with the hillslope system relates as a whole, the precipitation input into the basin with the surface runoff at the outlet (Rodriguz-Iturbe, 1979). This conception is imbedded in the work of Robert Horton (1945); an even greater accomplishment for hydrology than his specific laws in our opinion. He provided hydrologists and geomorphologists with a ladder of an ordering system, as a pattern which connects.

The whole network must embody a deep sense of regularity, not the trivial regularity of size, but the much deeper regularity in formal relations between the parts. This deeper regularity is contained-n two dimensions-in the Horton laws and in many other relationships derived subsequently for the two-dimensional characterization of the channel network. Rodriguz-Iturbe and Valdes (1979) addressed the basic question of relating the hydrologic response of a basin to its geomorphologic structure as described by Horton's laws. In that work an explicit linkage to the third dimension, the gravity, was not attempted. We feel it is mainly through this linkage that the interlocking between the network and the hillslope system will be accomplished. Hillslopes are the runoff producing elements which the network connects to accomplish the transformation of spatially distributed potential energy to kinetic energy at the outlet. The work thus executed is probably carried out through some general principle like minimum average work from the different parcels in the watershed. In fact energy expenditure may constitute the focal point for two opposing tendencies operating in the basin, minimum total work by the system which produces runoff and transports it to the outlet and minimum average work for each parcel of such a system. These two tendencies seem to be responsible for the tree-like branched structure of the drainage network. In the words of Stevens (1974), branching patterns are incredibly good compromises commanding the best of two worlds: shortness as well as directness.

It is our view that a structural regularity among the runoff producing areas of the hillslopes does exist. If this view is to be verified, then the third dimension, and more specifically the drops in elevation among the channel links of the network, should also be connected through some unifying principles. These unifying principles will permit the postulation of energy considerations dependent on the amount of runoff produced by the hillslopes which we think to be intimately linked with the drainage network. One step further in this general view will lead to the interlocking between the three dimensional structural regularity of the network and the runoff being produced by the hillslopes around the system of channels.

4. ON AN EMPIRICAL 'LAW' OF NETWORKS PARAMETERIZED BY ALTITUDES

Although the formulation in sec. 2 has proved to be useful in providing further understanding of the role of the network geometry in hydrologic response, the discussion in sec. 3 indicates the ways in which this formulation is still incomplete. More specifically, there are two compelling physical reasons which mandate that the formulation developed in sec. 2 be modified; (1) natural basins are three-dimensional and the altitudes play a dominant role in runoff generation and channel formation via transport of water and sediment under gravity, (2) apart from not having an unambiguous means of specifying flow distance, be it empirical or theoretical, the flow distance is part of the network which is being described and therefore is not entirely suitable as a reference parameter in quantifying networks, for example, as it appears in the width function $N(\cdot)$.

Heretofore the role of gravity has been implicitly accommodated for in the kernel $g(\cdot,\cdot)$. However the precise way in which this is done is obscure enough to be a continual source of discomfort at the foundational level. On physical grounds, it was discussed in sec. 3 that a search for structural regularity between the network and the hillslope system would most plausibly involve mass and energy considerations. Such ideas cannot be explored fruitfully without altitudes appearing in the geomorphologic descriptions in a fundamental way. Indeed as will be demonstrated below, a natural as well as useful reference parameter relative to which changes in the network structure can be described is **altitude.**

For the sake of convenience, we will henceforth assume that the basin outlet has zero altitude because it constitutes the reference point relative to which altitudes are measured. Let $H_1, H_2, \ldots$ denote the 'altitude' drops or heights of successive links in a network. These drops can be measured unambiguously from a topographic map by computing the differences (taken to be positive) between the two junctions comprising a link, or between a junction and a source. This brings us to the following postulate on the statistical regularity in link heights:

POSTULATE: In the absence of geologic controls, the networks that form in nature have statistically independent link heights having an exponential density function with parameter $\lambda(\cdot)$, which can vary with altitude.

Six river basins from different geographic regions, ranging in size from 1 sq. km to about 100

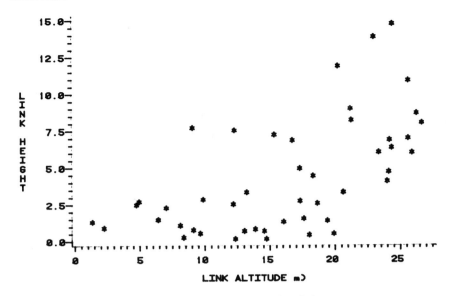

Figure 3   Heights versus Altitudes of Links for Goodwin
Creek, Sub-basin 13.

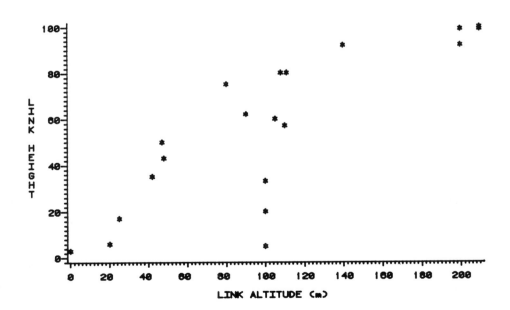

Figure 4   Heights versus Altitudes of Links for Morovis Basin.

sq. km were analyzed. The Goodwin Creek basin in Mississippi has relatively gentle slopes and crop land patterns, whereas the Mamon and Agua Fria basins in Venezuela are very mountainous; the first one with arid vegetation and the second having forested lands. In order to test homogeneity in the probability distributions of link heights, plots of link heights vs. link altitudes above the outlet for each of the six basins are shown in Figs. 2 through 7. Link altitude is taken to be the altitude of the upper junction of a link with respect to the basin outlet. As can be seen from these figures, departures from homogeneity appear to be present. This limited data indicates that the homogeneity issue in the probability distributions of link heights requires further empirical and theoretical investigations.

Figures 8 through 13 give plots of the empirical distribution functions (d.f.) of the link heights for each of the six basins. Exponential d.f. with constant parameters λ were fitted to each of these plots and these fits are also shown in Figs. 8 through 13. As one can see by inspection of the figures, the fits are remarkably good in the low and high values of the link heights, but not as good in the intermediate range. This observation in conjunction with Figs. 2 through 7 seems to suggest a separation into two intervals of constancy for the parameter $\lambda(h)$ of the exponential d.f. postulated above. However, the issue of nonhomogeneity in the link height d.f. requires a more careful study.

Table 1 lists the values of the mean link height λ^{-1} used in the exponential fits, the total network height H and the dimensionless number λH for each of the six basins. As can be seen from the table, the ratios of H over λ^{-1} lie between 5.4 to 9.6, while the magnitudes of these basins vary by an order of magnitude and the areas by two orders of magnitude. Considering the ambiguity in identifying first order channels and that the parameter λ may depend on the map scale (all the basins here are not drawn to a common scale, as shown in Table 1), the ratio seems fairly stable.

The observed empirical link distributions reported above certainly must depend on the map scale at which links are identified. We assume that there is a "self-similarity" over a range of map scales over which the distributional character in the links is maintained up to possible changes in the parameter λ. To fix ideas more concretely about the notion of a basin scale, let us consider an interval around an altitude h. Let the point functions $N(h)$ and $A(h)$ denote the number of links in this interval and the basin area density enclosed within this interval, respectively. It is assumed that the distribution of $N(h)$ and $A(h)$ differ by a shift and scale. The parameter $D_N(h)\,(L^{-1})$ scaling $A(h)$ to $N(h)$ through the ratio $N(h)/A(h)$ is called the "drainage scaling parameter". When the dimensions of the interval dh around the altitude h are very small, the ratio $D_N(h)$ would show significant fluctuations with small changes in the size of this interval. As the size increases one expects this ratio to gradually stabilize. The smallest size of the interval around h over which this stability occurs can be thought of as the representative elementary volume (REV) comprising the basin spatial scale. It is important to remark at this point that in our conception, "basin scale" is not tied to the size of a basin but rather to the size of an area within which a basin possesses an identifiable channel network and governs runoff generation from a system of hillslopes. For a general discussion about an REV, the reader may refer to Bear (1972). Typically one expects that the REV spans a range of dimensions over which the basin scale remains defined, as is generally true for a scale. In this sense an REV at the basin scale is analogous to, but much larger than, an REV at the hydrodynamic scale. It must be remarked here that the dimensions of an REV for the basin scale are also linked to the map scale, but we will not stop to explore this issue any further here.

A basin, at the basin scale, can be defined to be homogeneous with respect to scaling if the drainage scaling parameter does not vary with altitude, i.e., $D_N(h) \equiv D_N$. Similarly the

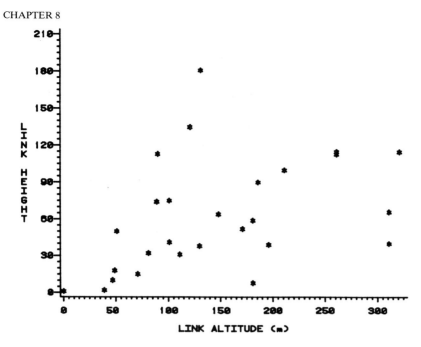

Figure 5   Heights versus Altitudes of Links for Unibon Basin.

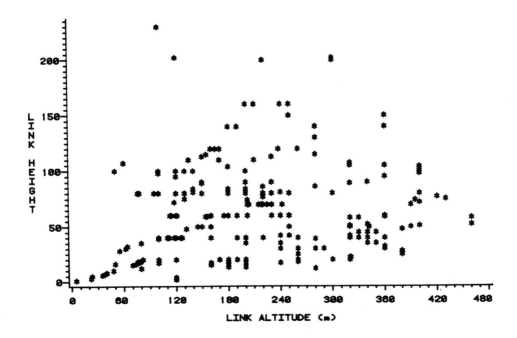

Figure 6   Heights versus Altitudes of Links for Agua Fria Basin.

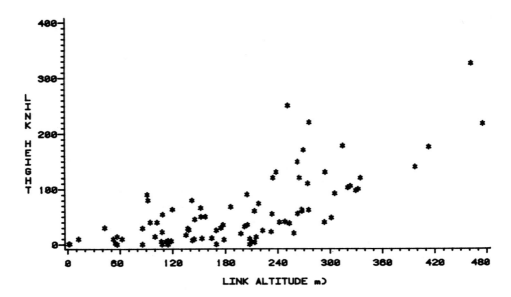

Figure 7 - Heights versus Altitudes of Links for Mamon Basin.

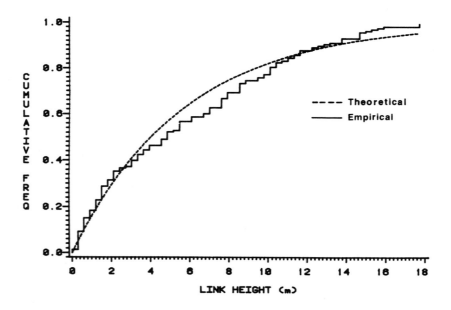

Figure 8   Empirical and Theoretical Distribution Functions
of Link Heights for Goodwin Creek Sub-basin 2.

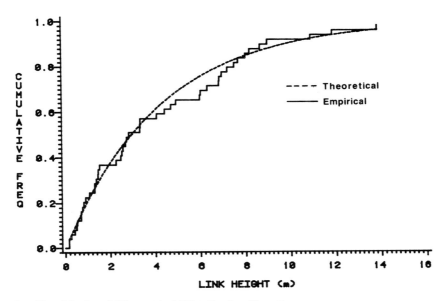

Figure 9   Empirical and Theoretical Distribution Functions
            of Link Heights for Goodwin Creek Sub-basin 13.

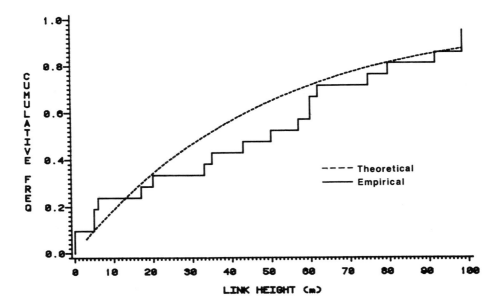

Figure 10   Empirical and Theoretical Distribution Functions
            of Link Heights for Morovis Basin.

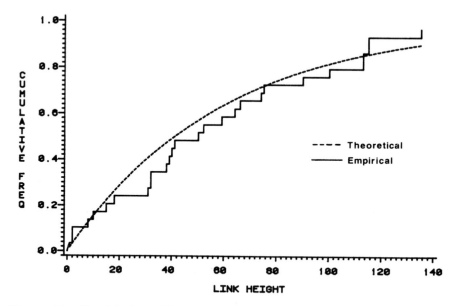

Figure 11   Empirical and Theoretical Distribution Functions
of Link Heights for Unibon Basin.

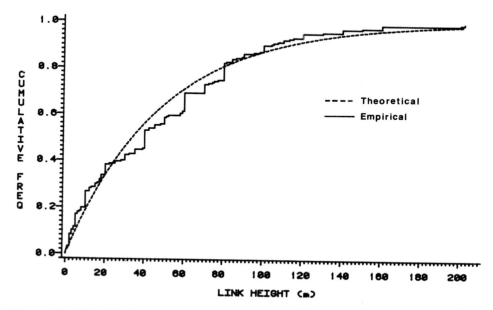

Figure 12   Empirical and Theoretical Distribution Functions
of Link Heights for Agua Fria Basin.

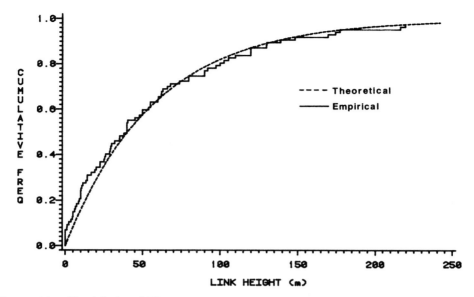

Figure 13   Empirical and Theoretical Distribution Functions
of Link Heights for Mamon Basin.

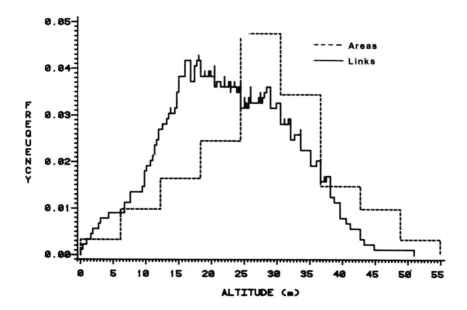

Figure 14   Altitudinal Variations in Link Concentration and
Area in Goodwin Creek Sub-basin 2.

TABLE 1. Geomorphologic and Scale Information for Six Basins

Basin Identi- fication	Network Height H(m)	Contour Interval/ Map Scale	Mean Link Height $\lambda^{-1}(m)$	Ratio λH	Area km^2	Magnitude M
Goodwin sub 2 MS	51.5	1.52 m 1:5000	5.8	8.9	17.7	77
Goodwin sub 13 MS	27.0	1.52 m 1:5000	4.3	6.3	1.3	25
Morovis Puerto Rico	260.0	10 m 1:20000	47.8	5.4	13.3	11
Unibon Puerto Rico	395.0	10 m 1:20000	59.0	6.7	23.3	15
Agua Fria Venezuela	470.0	20 m 1:25000	48.7	9.6	8.9	109
Mamon Venezuela	475.3		58.2	8.2	103.0	44

homogeneity with respect to shift can be defined. The reader may note that D_N is similar to (but not the same) as the drainage density introduced by Horton (1932). In order to qualitatively test the homogeneity assumption for the basins under study, histograms of basin areas versus altitude are shown for the Goodwin Creek basins, the two Puerto Rican basins and the Agua Fria basin in Figs. 14 through 18; such a plot could not be obtained for the Mamon basin. The link concentration function $N(h)$, denoting the number of links at altitude h, versus altitude is also plotted for the five basins in these figures. From these figures one can see that with changes in altitudes, $N(h)$ follows the same general pattern as the basin area histogram. In this connection, it may be noted that the link concentration function $N(h)$ can be expected to lie shifted to the left of the area-altitude histogram due to the fact that the external links do not grow up to the basin divide. A non-erodible belt lies between a source and the basin divide (Rodriguz-Iturbe, 1979). This effect is clearly seen in Figs. 14 and 15 for Goodwin Creek basin 2 and its sub-basin 13. This shift is about the same in both Figs. 14 and 15, thus supporting the homogeneity assumption with respect to shift. In Figs. 16 through 18 the shift is not that apparent because of the larger contour intervals and the map scales of these basins in comparison with those of the Goodwin Creek basin (see Table 1). These empirical observations support the assumption that these five basins are homogeneous at the basin scale with respect to shift and scaling. This homogeneity along with the empirical regularity in the link distribution structure lead to a preliminary but insightful analytical result about the fluctuations in runoff volume generated from the hillslopes along a channel network at the basin scale. This is discussed in the next section.

5. A MARKOV REPRESENTATION OF LINK CONCENTRATION AND SOME THEORETICAL CONSEQUENCES

The empirical results described in the preceding section suggest a natural theoretical representation of the evolution of network link concentration with respect to increases in elevation

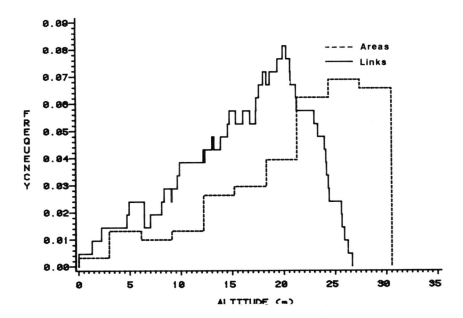

Figure 15   Altitudinal Variations in Link Concentration
and Area in Goodwin Creek Sub-basin 13.

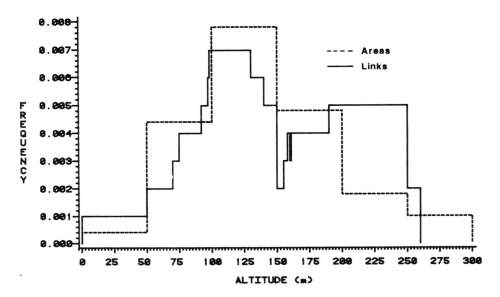

Figure 16   Altitudinal Variations in Link Concentration and
Area in Morovis Basin.

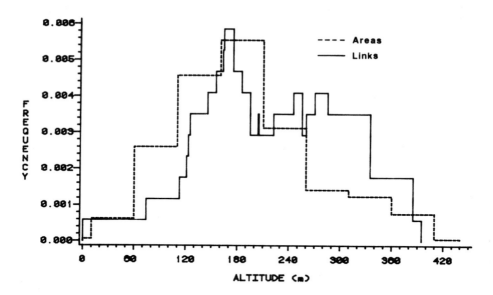

Figure 17   Altitudinal Variations in Link Concentration and
                 Area in Unibon Basin.

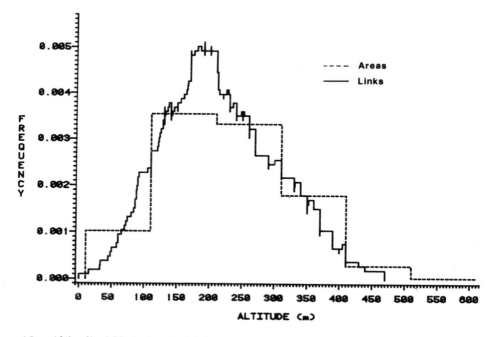

Figure 18   Altitudinal Variations in Link Concentration
                 and Area in Agua Fria Basin

within the basin. Recall that the basin altitude at the outlet is set equal to zero as a reference value.

The process $\{N(h)\}$ representing the numbers of links at altitudes $h \geq 0$ evolves by jumps of either one unit increase or one unit decrease corresponding to confluences or terminations of streams, respectively. The state space for the process is the set

$$S = \{0,1,2,...\} \tag{13}$$

and, moreover, the state 0 is an absorbing state in the sense that there are no further changes in $N(h)$ once state 0 is reached. Finally, the ground (or initial) value of $N(h)$ at $h = 0$ is unity,

$$N(0) \equiv 1. \tag{14}$$

Typical sample paths of the process $N(h)$ are represented in Figs. 14 through 18.

There are two important substructures embedded within the process $\{N(h)\}$ which can be singled out. The basin height is naturally partitioned by the network structure into a succession of heights $h_1 \leq h_2 \leq \cdots$ signaling the occurrence of a stream bifurcation or a stream termination within the network. In addition, let $h_0 = 0$ as mentioned earlier. The topological evolution of the two-dimensional network structure along the altitude is furnished by the values of the link concentration variables

$$N_k = N(h_k), \quad k = 0,1,2,... \tag{15}$$

Notice that, unlike the case of "link length", the notion of "link height" refers merely to the vertical (one dimensional) distance between two points and is apparently more stable with respect to scale fluctuation than is the planar (two dimensional) link length; i.e., altitudinal variation is small over regions of undecidability with regard to link termination events. The basin ensemble view will be adopted here for the theoretical representation of basin. While this point of view is common to much of the literature on quantitative geomorphology, the present approach permits the freedom to fix (via a map) the planar topological network structure by conditioning while still allowing for fluctuations in altitudes.

In view of the data analysis presented in Sec. 4, we will assume that the altitude over which a **single link** will evolve until bifurcation or termination is exponentially distributed with parameter $\lambda(h)$. In this connection it is important to make note that we are referring to the height of a single link randomly selected within the network and **not** to the differences $h_{i+1} - h_i, i = 0,1,...$ of the variables $h_0 \leq h_1 \leq h_2 \cdots$ in a plot of $N(h)$ versus h.

To first order in dh, the probability of a change in $N(h)$, by either a bifurcation or a termination in h to $h + dh$ is given by

$$P[\text{change in } N(h) \text{ in interval } (h, h + dh)] = N(h)\,\lambda(h)\,dh \tag{16}$$

under the assumptions that the $N(h)$ links have an equal chance $\lambda(h)dh$ of bifurcating or terminating in h to $h + dh$ independently of each other and that the probability of two or more changes at the same altitude is of smaller order than $(dh)^2$. It is also assumed that the conditional probability of an increase (bifurcation) at altitude h given that a change occurs is given by a parameter $p(h)$ and the probability of a decrease (termination) is $q(h) = 1 - p(h)$. To first order in dh the probability of an increase in $\{N(h)\}$ in h to $h + dh$ is given by

$$P[\text{increase in} N(h) \text{ in interval}(h, h + dh)] = p(h)\,\lambda(h)\,N(h)dh, \tag{17}$$

and the probability of a decrease in $\{N(h)\}$ in h to $h + dh$ is to first order in dh given by

$$P\,[\text{decrease in } N(h)\text{in interval } (h,h + dh)] = q(h)\,\lambda(h)\,N(h)dh \qquad (18)$$

The parameters $p(h)$ and $q(h)$ do not coincide with the bifurcation parameters of Shreve (1966) and need not be constant (see Mesa, unpublished dissertation, 1986). Under the assumptions that the network structure satisfies the topological randomness discussed above, independently of the altitude drops, and that the altitude over which $N(h)$ changes in state i is exponential with parameter $i\,\lambda(h)$, it follows that $\{N(h):h \geq 0\}$ is a (possibly non-homogeneous) Markov process with $N(0) = 1$ and absorbing state at 0. Let

$$f_{ij}(u,h) = \text{Prob}[N(h) = j \mid N(u) = i], \quad u \leq h \qquad (19)$$

denote the transition probability function of this Markov process. This function can be computed from the Kolmogorov forward equations, given by (Feller, 1968, Ch. 17),

$$\frac{\partial f_{ij}(u,h)}{\partial h} = (j - 1)\,\lambda(h)\,p(h)\,f_{i,j-1}(u,h) - j\,\lambda(h)\,f_{ij}(u,h) \qquad (20)$$

$$+ (j + 1)\,\lambda(h)\,q(h)\,f_{i,j+1}(u,h), \quad i \geq 0, \quad j \geq 1$$

$$\frac{\partial f_{io}(u,h)}{\partial h} = \lambda(h)\,q(h)\,f_{i1}(u,h), \quad i = 0,2,...$$

$$f_{oj}(h,h) = \delta_{oj}, \quad j = 0,1,2,... \quad h \geq 0.$$

Also, the conditional infinitesimal moments of the process $N(h)$ are given by

$$E\,\{N(h + \Delta h) - N(h) \mid N(h) = m\,\} = p(h)\lambda(h)m\,\Delta h - q(h)\lambda(h)m\,\Delta h = \alpha(h)m\,\Delta h$$

$$E\,\{[N(h + \Delta h) - N(h)]^2 \mid N(h) = m\,\} = p(h)\lambda(h)m\,\Delta h$$

$$+ q(h)\lambda(h)m\,\Delta h = \lambda(h)m\,\Delta h \qquad (21)$$

where

$$\alpha(h) = (p(h) - q(h))\lambda(h).$$

It may be recalled from our discussion in Sec. 4, that the distributional character of the link heights must remain invariant over a range of scales. Since the observed path structure of $\{N(h)\}$ certainly depends on the map scale at which links are identified, we assume that at the basin scale, as defined in Sec. 4, the statistical structure of $\{N(h)\}$ is Markov; the parameters possibly being dependent on the map scale. Certain aspects of the moment structure of $N(h)$ are given in Bailey (1964). The Markov structure of $N(h)$ will be used below for illustrating the nature of fluctuations in runoff volume generated from a network.

The runoff generation from any surface of a basin depends in a complex manner on a host of physical processes, e.g., precipitation, infiltration, evapotranspiration, evaporation, etc. These processes, in turn, are governed by climatic conditions, basin terrain, geology, etc. However, we discussed in Sec. 3 that a minimum amount of flow is necessary to maintain a channel. This means that the presence of a link, in particular an external link, at some elevation represents the appearance of a minimum runoff from the hillslopes of that link over a long time span. This leads us to speculate that the runoff volume $R(dh)$ in an increment of elevation dh can be expressed as

$$R(dh) = \bar{\mu}(h) \, N(h) \, dh \, , \tag{22}$$

where $\bar{\mu}(h)$ denotes the runoff volume per link per unit elevation. The identity given by Eqn. (22) must be interpreted as an average runoff volume over some long time span, e.g., a season or a year, rather than over the time scale of a single rainfall-runoff event. At the time scale of single rainfall-runoff events, the runoff volume per unit elevation of a link must be viewed as a (stochastic) time dependent function, $\mu(h,t)$. Then $\bar{\mu}(h)$ can be regarded as the time average of $\mu(h,t)$, via the law of large numbers, as $t \to \infty$. However we need not get into such modeling details at present, since our objective for now is only to illustrate how the ensemble view for a network can lead to an understanding of the "laws of fluctuations" governing the process of runoff generation from hillslopes along the network.

The total runoff R from a basin can be obtained by integrating Eqn. (22) over the altitudes

$$R = \int_0^\infty R(dh) = \int_0^\infty \bar{\mu}(h) \, N(h) \, dh \tag{23}$$

Since a basin has only finite altitude, Eqn. (23) remains well defined but for notational convenience the upper limit of integration in this equation is taken to be infinity. In homogeneous basins at the basin scale, since $D_N(h) \equiv D_N$, it appears reasonable to assume that $\mu(h)$ will also be a constant, say $\bar{\mu}$, independent of altitude. In this case Eqn. (23) can be written as

$$R = \bar{\mu} \int_0^\infty N(h) \, dh \tag{24}$$

As discussed earlier in this section, in the graph of $N(h)$ versus h the network events corresponding to bifurcation and termination of channels produce a natural partitioning of the basin height by $0 \equiv h_0 \le h_1 \le h_2, \cdots$. In the interval $[h_i, h_{i+1}]$, the function $N(h) \equiv N(h_i) = N_i$. Therefore R in Eqn. (24) can be expressed as being proportional to the area under $\{N(h)\}$,

$$R = \bar{\mu} \int_0^\infty N(h) \, dh = \bar{\mu} \sum_{i=0}^{2M-2} N_i (h_{i+1} - h_i), \tag{25}$$

where M is the basin magnitude; recall that there are a total of $2M-1$ links in a basin of magnitude M. Under the assumption that $\lambda(h)$ is uniform with respect to altitudes, i.e., $\lambda(h) \equiv \lambda$, Eqn. (25) leads to the following result:

> R is conditionally, given the network magnitude, gamma distributed with parameters $\lambda/\bar{\mu}$ and $2M-1$.

To see the reason for this first simply note that given the magnitude M and the values $N_1, N_2, \ldots$, the drop $\Delta h_i = h_{i+1} - h_i$ is exponentially distributed with parameter λN_i. This follows from the Markov property or the lack of memory property of an exponential distribution (Feller, 1968, p. 329) and the assumption that the links bifurcate and terminate statistically independently of one another. Therefore $N_i \Delta h_i$ is conditionally exponential with parameter λ. Since the sum of $2M - 1$ independent and identically distributed exponentials is a gamma and since this gamma density only depends on M and λ and not on the values $N_1, N_2, \ldots$, etc., summing with respect to the joint distribution of $N_1, N_2, \ldots$, gives the above results.

This result provides a very interesting structural regularity in fluctuations in the spatial distribution of runoff generated from hillslopes along a network in an homogeneous basin. It says that the total runoff generated over a long time span by any sub-basin associated with each link of a basin, has a gamma distribution with parameters $\lambda / \overline{\mu}$ and 2(link magnitude) - 1. The parameters λ, $\overline{\mu}$, M, etc. governing this fluctuation are the large scale parameters measurable and meaningful only at the basin scale. In this sense this result significantly generalizes the topological interpretation of the total runoff associated with a link, given by 2(link magnitude) - 1. When the parameter $\lambda(h)$ varies with altitude, the above result does not hold.

6. CLOSING REMARKS

The developments outlined in secs. 4 and 5 illustrate a framework for further explorations toward constructing a physically rigorous theory of the runoff process at the basin scale. We think that the next natural direction within this framework is to explore the distribution of potential energy in a basin along similar lines as developed in sec. 5 for runoff generation. In searching for structural regularities between fluctuations in potential energy, in kinetic energy and in losses via channels' morphology, it will be necessary to quantify the channel flow paths parameterized by altitudes. At present this seems to be a difficult problem. Only after a suitable transformation between altitudes and flow paths is established and empirically tested, will it be possible to compute flows from a basin as well as to begin to address the fundamentally important problem of hydrologic similarity among basins. Another important problem in searching for the physics of transformations of potential energy to kinetic energy appears to be tied to an understanding of the connections between the basin time scale and the basin spatial scale.

The problem of hydrologic similarity in runoff also requires that the role of climate, for example, via rainfall, on runoff be investigated. At present it seems most fruitful to understand the structure of fluctuations in runoff introduced by a basin separately from fluctuations in runoff introduced by the climate. A coupling of climate induced and basin induced fluctuations on runoff can be undertaken after each of these two areas has sufficiently progressed.

ACKNOWLEDGMENT

Gratitude is expressed to Oscar Mesa for providing invaluable assistance in data analysis for this work. This research was supported by Grant 21078-GS from the Army Research Office. The first two authors gratefully acknowledge the hospitality and support extended by the Utah Water Research Laboratory at Utah State University during the course of this study.

REFERENCES

Bailey, N.T.J., 1964, *The Elements of Stochastic Processes With Applications to Natural Sciences*, Wiley, New York.

Bear, J., 1972, *Dynamics of Fluids in Porous Media*, American Elsevier, New York.

Beven, K., 1986, "Runoff Production and Flood Frequency in Catchments of Order n: An Alternative Approach", (this issue).

Dunne, T., 1978, "Field Studies of Hillslope Flow Processes", Ch. 7 in *Hillslope Hydrology*, (ed.) M. Kirkby, J. Wiley & Sons.

Dunne, T., 1982, "Models of Runoff Processes and Their Significance", In: T. Fiering (Chm), *Scientific Basis of Water Resources Management*, National Academy Press.

Eagleson, P.E., 1970, *Dynamic Hydrology*, McGraw-Hill, New York.

Feller, M., 1968, *An Introduction to Probability Theory and Its Applications*, Vol. I, Wiley, New York.

Gupta, V.K., and E. Waymire, 1983, "On the Formulation of an Analytical Approach to Hydrologic Response and Similarity at the Basin Scale", *J. Hydrol.*, 65(1/3), pp. 95-123.

Gupta, V.K., E. Waymire, and C.T. Wang, 1980, "A Representation of an Instantaneous Unit Hydrograph from Geomorphology", *Water Resources Research*, 16(5), pp. 855-862.

Hewlett, J.D., 1961, "Watershed Management", In: Report for 1961 Southeastern Forest Experimental Station, U.S. Forestry Service, Asheville, North Carolina.

Hewlett, J.D., and A.R. Hibbert, 1967, "Factors Affecting the Response of Small Watersheds to Precipitation in Humid Areas", In: W.E. Sopper and H.W. Lull (eds), *Forest Hydrology*, Pergamon, Oxford, pp. 275-290.

Horton, R.E., 1932, "Drainage Basin Characteristics", *EOS Trans., AGU*, Vol. 13, pp. 350-361.

Horton, R.E., 1945, "Erosional Development of Streams and Their Drainage Basins: Hydrophysical Approach to Quantitative Morphology", *Bull. Geol. Soc. Amer.*, Vol. 56, pp. 275-370.

Kirkby, M.J., 1976, "Tests of the Random Network Model, and Its Application to Basin Hydrology", *Earth Surf. Proc.*, Vol. 1, pp. 197-212.

Kirkby, M.J., 1978, *Hillslope Hydrology*, John Wiley, New York.

Kirkby, M.J., 1986, "A Runoff Simulation Model Based on Hillslope Topography", (this issue).

Mesa, O., 1982, On an Analytical Approach for Coupling Hydrologic Response and Geomorphology. Unpublished M.S. Thesis, Department of Civil Engineering, University of Mississippi.

Mesa, O., and E.R. Mifflin, 1986, "On the Relative Role of Hillslope and Network Geometry in Hydrologic Response", (this issue).

Pilgrim, D., 1983, "Some Problems in Transferring Hydrological Relationships Between Small and Large Drainage Basins and Between Regions", J. Hydrol., 65(1/3), pp. 49-72.

Rodriguez-Iturbe, I., 1979, "The Physical Basis of Surface Water Hydrology: Climate and Watershed Development", Paper presented at the Fall AGU Meeting, San Francisco.

Rodríguez-Iturbe, I., and J.B. Valdes, 1979, "The Geomorphic Structure of Hydrologic Response", Water Resources Research, Vol. 15, No. 6, pp. 1409-1420.

Rodríguez-Iturbe, I., M. Gonzalez-Sanabria, and R. Bras, 1982, "A Geomorphoclimatic Theory of the Instantaneous Unit Hydrograph", Water Resources Research, Vol. 18, No. 4. pp. 877-886.

Shreve, R.L., 1966, "Statistical Law of Stream Numbers", J. Geol., 74, pp. 17-37.

Shreve, R.L., 1969, "Stream Lengths and Basin Areas in Topologically Random Channel Networks", J. Geol., 77, pp. 397-414.

Shreve, R.L., 1974, "Variation of Mainstream Length with Basin Area in River Networks", Water Resources Research, Vol. 10, No. 6, pp. 1167-1177.

Stevens, P.S., 1974, Patterns in Nature, Little, Brown & Co., Boston.

Troutman, B.M. and M. R. Karlinger, 1984, "On the Expected Width Function for Topologically Random Channel Networks", J. of Appl. Prob., Vol. 22, pp. 836-849.

Troutman, B.M. and M. R. Karlinger, 1985, "Unit Hydrograph Approximations Assuming Linear Flow Through Topologically Random Channel Networks", Water Resources Research, Vol. 21, No. 5, pp. 743-754.

Troutman, B.M. and M.R. Karlinger, 1986, "Averaging Properties of Channel Networks Using Methods in Stochastic Branching Theory", (this issue).

9

AVERAGING PROPERTIES OF CHANNEL NETWORKS USING METHODS IN STOCHASTIC BRANCHING THEORY

Brent M. Troutman
Michael R. Karlinger

ABSTRACT

Methods in branching theory are used to average properties of channel networks, resulting in expressions for the instantaneous unit hydrograph (IUH) in terms of fundamental network characteristics (Z, α, β), where α parameterizes the link (channel segment) length distribution and β is a vector of hydraulic parameters. Several possibilities for Z are considered, including N, (N, D), (N, M), $\tilde{D}$, and $(N, \tilde{D})$, where N is magnitude (number of first-order streams), D is topological diameter, M is order, and $\tilde{D}$ is mainstream length. Linear routing schemes, including translation, diffusion, and general linear routing, are used, and it is demonstrated that translation routing leads to an IUH identical to that obtained by use of the width function, where, for a given distance x, the width of a network is defined to be the number of links some point of which lies at channel distance x from the outlet (analogous to population size in branching theory).

The IUH is taken to be the conditional expectation of actual basin response given Z, and it is derived based on assumptions that the links are independent and identically distributed random variables and that the network is a member of a topologically random population. Uncertainty in use of this expectation to approximate actual basin response is given by the corresponding conditional variance.

Asymptotic (for large N, D, and $\tilde{D}$) results are available for several cases. For example, when $Z = N$ asymptotic considerations lead to a Weibull probability density function for the IUH for all linear routing schemes, with only a single parameter depending on the particular routing method.

A simulation study compares different possibilities for Z in terms of ability to predict actual IUH characteristics. These characteristics include peak and time to peak.

V. K. Gupta et al. (eds.), Scale Problems in Hydrology, 185–216.

1. INTRODUCTION

In this paper we explore in detail the application of methods in stochastic branching theory to defining the hydrologic response of a drainage basin in terms of a few fundamental basin characteristics. This paper builds upon and generalizes the work presented in Troutman and Karlinger (1984, 1985) and Karlinger and Troutman (1985). The paper by Karlinger and Troutman (1985) applies the theory developed in the other two papers to two actual drainage basins Blue Creek in Alabama and Cane Branch in Kentucky. Hydrograph approximations based on fundamental basin characteristics agree well with actual hydrographs based on detailed modeling of the basins.

The use of fundamental basin characteristics to approximate actual basin response involves an averaging of basin properties, both geomorphologic properties and hydraulic properties. The approach in the present work is to use probabilistic models as a means of performing this averaging. This approach has been an area of intense research in recent years. Much of this latest research has built upon the methods presented in works by Rodriguéz-Iturbe and Valdés (1979) and Gupta, Waymire, and Wang (1980). In the present paper, we are in particular interested in applications of stochastic branching theory to channel networks. Principles of branching theory also were exploited by Gupta and Waymire (1983).

The general reasoning behind using stochastic models to approximate a function of many variables (actual runoff) with a function of a few variables (approximate runoff determined using fundamental basin characteristics) is discussed by Troutman and Karlinger (1985). They also discuss how conditional expectations may be used to define the functional form of approximate runoff. This approach also is taken in this paper. The stochastic structure to be applied to drainage basins is determined by the assumption of topological randomness [see, for example, Shreve (1967) and Smart (1972)], which involves taking both the channel segment lengths and the topological configuration (or, the way in which the channel segments fit together) to be random quantities. We also introduce the notion of letting the hydraulic properties be random variables, but most of the analysis is based on an assumption that these properties are constant.

In the present paper, as in our previous works, we are concerned only with movement of water on the surface through a channel network. Infiltration, subsurface flow, and overland flow are not considered here. It is assumed that we have a certain quantity of water that has been deposited in an existing channel network, and we wish to examine how this water is routed to the outlet of the drainage basin. In particular, we are interested in basin outflow that results from depositing a unit amount of water instantaneously and uniformly into a channel network [or the instantaneous unit hydrograph (IUH)]. It is assumed that the response for a more complicated rainfall pattern can be obtained by the usual convolution technique. This IUH approach also was used by Rodriguéz-Iturbe and Valdés (1979) and Gupta, Waymire, and Wang (1980). Because of their mathematical tractability, linear-routing schemes are used to route water through the channel networks in the present paper.

2. DEFINITIONS

In this section we give some definitions and review some basic results for topologically random channel networks. Many of the basic notions and results may be found in papers of Shreve (1966, 1967, 1969, 1974) and Smart (1968, 1972, 1978).

The **outlet** of a network is the point farthest downstream, **sources** of the network are points farthest upstream, and a point at which two channels join to form one channel is called a **junction.** The assumption is made that no more than two channels combine at a single junction. **Exterior links** are defined to be segments of the channel network between a source

and the first junction downstream, and the **interior links** are segments between two successive junctions or between the outlet and the first junction upstream.

In applying branching theory to the evolution of random channel networks, it also is necessary to speak of the **bifurcation probability,** p, and the **termination probability,**
$q = 1-p$, associated with an individual link. If we view the evolution of the network as originating at the outlet and proceeding upstream (analogous to a population model in branching
theory, as discussed later), a given link may either terminate, in which case it becomes an
external link, or bifurcate, in which case it becomes an internal link.

The **magnitude,** N, of a network is the number of sources, or exterior links, in the network.
The distinguishing feature of **topologically random networks** is that, given the magnitude
$N = n$, all distinct configurations having n sources are assumed to be equally likely.

The **topological** distance of a given source from the outlet of a network is the number of
links (one of which is exterior) in the path joining these two points, and the **diameter,** D, of
the network is the maximum (over all the sources) such distance. The **actual** distance of a
source from the outlet of a network is the sum of the link lengths in the path joining these
two points, and the **mainstream length,** $\tilde{D}$, of the network is the maximum such distance.
The diameter can be considered to be a topological mainstream length. The **order** of a link
is defined as follows: All exterior links have order 1, and if two links of orders m_1, m_2 join,
the order of the next downstream link is $\max(m_1, m_2)$ if $m_1 \neq m_2$ and is m_1+1 if $m_1 = m_2$.
The order, M, of a network is defined to be the order of the link between the outlet and the
first junction upstream. This is the so-called Strahler ordering system.

In working with topologically random networks, it often is necessary to compute the number
of topologically distinct configurations for networks having given values for N, (N,D), and
(N,M). Denote these numbers by p_n, q_{dn}, and r_{mn}, respectively. Methods for computing
these quantities first were investigated by Shreve (1966, 1967, 1969, 1974). Troutman and
Karlinger (1984) deal with ways of computing the generating function of p_n q_{dn}, and r_{mn}.

$$\overline{p}(s) = \sum_{n=0}^{\infty} 4^{-n} \ p_n \ s^n.$$

Troutman and Karlinger (1984, Theorem 4.1) give

$$\overline{p}(s) = [1 - (1-s)^{1/2}]/2, \tag{1}$$

and they give nonlinear difference equations which may be used to obtain generating functions for q_{dn} and r_{mn}. It is readily shown that p_n is given by

$$p_n = (2n - 1)^{-1} \binom{2n-1}{n} \tag{2}$$

For a given network, assume that we use the following algorithm for numbering the links
(Shreve, 1967): The network is traversed by starting at the outlet, turning left at junctions,
reversing direction at sources, and numbering the links consecutively as they are traversed the
first time. The sequence of variables $\{X_i\}$ will be used to denote link lengths, with the j-th
link (according to the numbering scheme given above) being assigned length X_j.

In this paper it will be useful to define an integer-valued random variable, T, the value of
which will completely specify the **topological configuration** of the network under consideration. For example, we may let $T = 1, 2, 3, 4$ correspond to networks with 1 source, 2
sources, 3 sources with one configuration, and 3 sources with another configuration, respectively. (There are two topologically distinct configurations for networks with 3 sources; see
Figure 1 in Troutman and Karlinger (1985)). It is unimportant in this paper exactly how this

association is made for larger networks, so we state here simply that it can be done. We also note that N, M, and D are functions of T; that is, given T, we automatically know N, M, and D.

3. APPROXIMATING RUNOFF RESPONSE USING FUNDAMENTAL NETWORK CHARACTERISTICS

3.1. Use of the Conditional Expectation

In this section we describe in general terms how the IUH for a basin may be approximated using only fundamental network characteristics. (For more details, see Troutman and Karlinger, 1985.) We first assume that for channel routing purposes a drainage basin may be characterized completely by the following properties:

> (1) Topological configuration, T;
>
> (2) lengths of the individual links,
>
> $X_1, X_2, X_3, \cdots$;
>
> (3) hydraulic properties of the individual links,
>
> $B_1, B_2, B_3, \cdots$.

The set of hydraulic properties for the i-th link, B_i, may be a vector consisting of several properties used to route water through the channel segment. We denote a vector consisting of all the link lengths by X and a vector consisting of all the hydraulic properties by B. Thus, given T, X, and B for a basin, we could route (probably numerically) a given inflow through the network to obtain the hydrograph at the outlet. We have not yet placed any restrictions on what routing technique is to be employed. We will denote basin response to an instantaneous unit input (the IUH) by

$$\text{IUH} = I(t; T, X, B), \tag{3}$$

showing explicitly the dependence on basin properties; t is time.

Our goal in approximating the IUH is to avoid having to completely specify all the basin properties T, X, and B. We wish to define a vector, say $Z*$, consisting of a few fundamental properties that are believed to be important in determining the IUH and that may be determined much more easily than are T, X, and B. We then must find a function, say $\bar{I}$, of these fundamental properties that yields a good approximation to the IUH:

$$\bar{I}(t; Z*) \doteq I(t; T, X, B). \tag{4}$$

There are many ways to obtain $\bar{I}$ once a vector $Z*$ has been selected. One very general approach is to postulate a probability model that is assumed to govern (T, X, B), and then to derive $\bar{I}$ based on operations from probability theory. This was done, for example, in recent works by Rodriguéz-Iturbe and Valdés (1979) and Gupta, Waymire and Wang (1980). In this paper, we follow the same general approach; the probability model governing (T, X, B) consists of a minimum of basic assumptions, which we state explicitly in later sections. We begin by specifying a vector, Z, of fundamental basin properties (as we shall see, Z makes up part of $Z*$), and we then use the conditional expectation of $I(t; T, X, B)$, given $Z = z$, as the approximate IUH, $\bar{I}$ in (4). Here, upper case Z denotes a random variable, and lower case z denotes an observed value of Z. The expectations is taken with respect ot the probability distribution of (T, X, B), which we shall assume is parameterized by a fixed vector, λ. Thus, the expectation is a function of both z and λ :

$$\bar{I}(t;z;\lambda) = E[I(t;T,X,B)|Z = z]. \qquad (5)$$

The vector $Z*$ in (4) is thus

$$Z* = (Z;\lambda).$$

The parameter λ is estimated by $\hat{\lambda}$, say, so we actually use $\bar{I}(t;z;\hat{\lambda})$ as the IUH

We can, in principle, obtain the approximation in (5) for any routing scheme, for any distribution of (T,X,B) and for any Z we choose to use. The expression is, in general, of limited value, however, because computing the given conditional expectation may be at least as difficult as performing the routing through the basin. In this paper we make several simplifying, but physically realistic, assumptions that lead to elegant ways of obtaining (5). First, it is going to be assumed that the hydraulic parameters B_i are identically equal to a single "effective" value β that is nonrandom and constant for all links. Second, T and X will be assumed to be independent, and the probability distribution of X_i is parameterized by α. Also, we assume that the basin is a member of a topologically random population, which determines the distribution of T; this distribution has no parameters. Thus $\lambda = (\alpha,\beta)$. Finally, we are going to use only linear routing schemes. (See, for example, Harley, 1967; Keefer and McQuivey, 1974.)

3.2. Fundamental Basin Properties

In this paper, we consider several possibilities for the fundamental basin properties, Z :

(a) $Z = N$,

(b) $Z = (N,D)$,

(c) $Z = (N,M)$,

(d) $Z = \tilde{D}$,

(e) $Z = (N,\tilde{D})$.

Note that the properties, Z, in (a), (b), and (c) are a function only of T, but not of X or B; that is, these properties can be determined from the topological configuration of the basin without knowing anything about the properties of the links themselves.

3.3. Comparison with Previous Work

The connection between the approach taken in this paper and that of other authors is discussed in detail by Troutman and Karlinger (1985). We briefly summarize some of the connections here.

Rodriguéz-Iturbe and Valdés (1979) also approximate I in (3) by an averaged hydrograph, $\bar{I}$ in (4), which is a function of fundamental basin characteristics. Their averaging process, however, does not consist of explicit evaluation of a conditional expectation, as we have done in (5). Rather, they have modeled movement through a network as a Markov chain with state space consisting of stream orders. Their vector of fundamental basin characteristics is

$$Z* = (M,R_A,R_B,R_L,\bar{L},V) \qquad (7)$$

where M is order, R_A is the area ratio, R_B is the bifurcation ratio, R_L is the length ratio, $\overline{L}$ is the mean first-order stream length, and V is a hydraulic parameter (velocity). Gupta, Waymire, and Wang (1980) generalized the approach of Rodriguéz-Iturbe and Valdés (1979), though the basic line of reasoning is similar.

Of central importance in this area of research is the assumed nature of the hydrologic response of individual channel segments. The two papers discussed above consider the "holding time" distributions for a channel segment. Wang, Gupta, and Waymire (1981) discuss in detail nonlinearity introduced by allowing the mean holding time to depend on rainfall intensity. Kirshen and Bras (1983), still within the framework of the IUH theory developed by Rodriguéz-Iturbe and Valdés (1979) and Gupta, Waymire, and Wang (1980), use an explicit solution for the linearized equations of motion to determine the response of individual segments. The linear routing scheme they utilize is identical to what we have chosen to label "general linear routing" in this paper.

Gupta and Waymire (1983) consider some properties of topologically random networks for which the link lengths are assumed to be constant. Their vector of fundamental basin characteristics is, using notation in the present paper, $Z = (N, D, L_1, L_2, ..., L_D)$, where N is magnitude, D is diameter, and, if the basin is taken to be a headward-growth branching process, L_i is the number of "offspring" (or links) at the i-th stage of branching. It will be seen that L_i is a version of the "width function" $L(x)$ for a basin (see Kirkby, 1976; Troutman and Karlinger, 1984, 1985). They also, as we have done, obtain a representation for the IUH by conditioning on Z.

The development of Gupta and Waymire used some ideas from statistical mechanics. (See also Lienhard, 1964.) One interesting result is that the form of the IUH is a function of the nature of constraints placed on the watershed system, although, as they point out, the physical justification for using these constraints is not clear. One particular set of constraints leads to a Weibull probability density (eq. 20 in Lienhard), which is precisely the form of an asymptotic result given in this paper.

3.4. Uncertainty in the Approximate IUH

Uncertainty in using $\overline{I}(t;Z*)$ to approximate $I(t;T,X,B)$ may be measured in a number of ways. One straightforward measure is the mean-squared error, which is a function, denoted by ξ, of t and λ:

$$\xi(t;\lambda) = E\{[I(t;T,X,B) - \overline{I}(t;Z*)]^2\} \qquad (8)$$

If $\overline{I}$ is the conditional expectation given in (5), this mean-squared error is

$$\xi(t;\lambda) = E\{Var[I(t;T,X,B)|Z]\} \qquad (9)$$

where the expectation here is with respect to the distribution of Z. This mean-squared error provides a convenient method for comparing two different sets of fundamental basin characteristics, say Z_1 and Z_2. If $\xi_1(t;\lambda)$ is (9) using Z_1 and $\xi_2(t;\lambda)$ is (9) using Z_2, then $\xi_1(t;\lambda) \leq \xi_2(t;\lambda)$ for all t would indicate that use of Z_1 gives a uniformly better approximation to the actual IUH, and vice versa.

Also, uncertainty in various functionals of the IUH may be of more interest than uncertainty for each t. For example, let T_p and $\overline{T}_p$ be the actual and approximate times to peak, respectively. The $E[(\overline{T}_p - T_p)^2]$ would measure uncertainty in this quantity.

It also may be desirable to restrict the set of basins for which a given measure of uncertainty is computed. For example, the expectation in (8) is over all possible drainage basins, complex and simple. Combining such a large range of basins may lead to confusion in interpretation of a mean-squared error. Suppose, on the other hand, that we restrict the set of basins under consideration to those with exactly n first-order streams. The expectation in (8) then would become a conditional expectation, given $N = n$.

4. LINEAR CHANNEL ROUTING

4.1. Linearization of the One-dimensional Flow Equations

In the remainder of this paper we shall be concerned with one-dimensional routing of flow in channels that are presumed to be wide and rectangular and in which frictional effects are assumed to follow the Chezy law. We assume further that the equations of motion have been linearized (Harley, 1967; Dooge and Harley, 1967; Keefer and McQuivey, 1974; Kirshen and Bras, 1983), so that discharge at the downstream end of a channel segment may be obtained by computing the convolution of the inflow at the upstream end of the segment with the response to a delta-function inflow. The reason for restricting ourselves to linear routing will be discussed later. We point out, however, that effects of nonlinearity could be introduced by the multiple linearization technique of Keefer and McQuivey (1974).

We consider the following situation: Rain is falling uniformly over a drainage basin for a period of time until an equilibrium discharge q_o (units are $t^{-1}l^3$) is reached. Assume that the lateral inflow rate (which is a function of the rainfall rate) at equilibrium is constant at every point in the network and is equal to r_o $(t^{-1}l^{-1}l^3)$; this is essentially an assumption of constant drainage density. The equilibrium discharge q_o will be a function of r_o, and for a given point in the network, will be proportional to the total channel length upstream from this point. Next, assume that an instantaneous burst of rain occurs uniformly over the basin at time $t = 0$ and is deposited immediately into the channel network; denote by v $(l^{-1}l^3)$ the volume of water per unit channel length that is deposited in the network, and again assume that this is constant at all points in the network. The response of the basin to this instantaneous burst is the IUH

The one-dimensional equations of motion for unsteady flow in a wide rectangular open channel are:

$$\frac{\partial q}{\partial w} + B \frac{\partial y}{\partial t} = r \tag{10}$$

and

$$\frac{\partial y}{\partial w} + \frac{u}{g} \frac{\partial u}{\partial w} + \frac{1}{g} \frac{\partial u}{\partial t} = S_0 - S_f - \frac{u}{Bgy} r$$

where g = gravitational acceleration $(1t^{-2})$,

t = time (t),

w = distance along the channel (l),

$q = q(w,t)$ = discharge $(t^{-1}l^3)$,

$y = y(w,t)$ = depth of flow (l),

$u = u(w,t)$ = velocity of flow $(t^{-1}l)$,

B = surface width of flow (l),

$r = r(w,t) = $ lateral inflow per unit length $(t^{-1}l^2)$,

$S_o = $ slope of channel bottom, and

$S_f = $ friction slope.

Frictional effects are assumed to follow the Chezy law:

$$S_f = u^2/(C^2 y), \tag{12}$$

where C is the Chezy coefficient. Putting $u = q/(By)$ in (11) and using (10) lead to

$$(B^2 gy^3 - q^2)\frac{\partial y}{\partial w} + 2qy\frac{\partial q}{\partial w} + By^2\frac{\partial q}{\partial t} = B^2 gy^3(S_o - \frac{q^2}{B^2 C^2 y^3}) \tag{13}$$

We now are going to separate the actual discharge q into two components; this is consistent with the scenario described previously. First, we let $q_o(w)$ and $y_o(w)$ denote the equilibrium discharge and depth before the instantaneous burst of rain. Note that q_o and y_o do not depend on time but may depend on distance. We then let upper case $Q(w,t)$ and $Y(w,t)$ be the deviations of actual depth and discharge from the equilibrium values. Thus,

$$y(w,t) = y_o(w) + Y(w,t) \tag{14a}$$

and

$$q(w,t) = q_o(w) + Q(w,t). \tag{14b}$$

The equilibrium q_o and y_o will obey

$$\frac{\partial q_o}{\partial w} = r_o \tag{15}$$

and

$$(B^2 gy_o^3 - q_o^2)\frac{\partial y_o}{\partial w} + 2q_o y_o\frac{\partial q_o}{\partial w} = B^2 gy_o^3(S_o - \frac{q_o^2}{B^2 C^2 y_o^3}), \tag{16}$$

from (10) and (13). To obtain differential equations for the deviations Q and Y, it is assumed, first, that these deviations are very small compared to q_o and y_o, and, second, that $\partial y_o/\partial w$ and $\partial q_o/\partial w$ are very small compared to q_o and y_o. The second assumption will hold if the equilibrium values do not exhibit too much change in a particular channel segment. Substituting (14a) and (14b) into (13), and ignoring terms in Y, Q, $\partial y_o/\partial w$, and $\partial q_o/\partial w$ of order greater than first gives

$$(B^2 y_o^3 g - q_o^2)\left[\frac{\partial y_o}{\partial w} + \frac{\partial Y}{\partial w}\right]$$

$$+ 2q_o y_o\left[\frac{\partial q_o}{\partial w} + \frac{\partial Q}{\partial w}\right] + y_o^2 B\frac{\partial Q}{\partial t}$$

$$= B^2 gS_o(y_o^3 + 3y_o^2 Y) - \frac{g}{C^2}(q_o^2 + 2q_o Q).$$

Subtracting (16) leaves

$$(B^2 y_o^3 g - q_o^2)\frac{\partial Y}{\partial w} + 2q_o y_o\frac{\partial Q}{\partial w} + y_o^2 B\frac{\partial Q}{\partial t} \tag{17}$$

$$= 3B^2 gS_o y_o^2 Y - \frac{2gq_o Q}{C^2}.$$

Similarly, we get

$$\frac{\partial Q}{\partial w} + B \frac{\partial Y}{\partial t} = R ,$$ (18)

where $R = r - r_o$. We next differentiate (17) with respect to t and (18) with respect to w and combine to obtain

$$(B^2 y_o^3 \ g \ - q_o^2) B^{-1} \left[\frac{\partial R}{\partial w} - \frac{\partial^2 Q}{\partial w^2} \right] + 2 q_o \ y_o \ \frac{\partial^2 Q}{\partial w \, \partial t}$$

$$+ \ y_o^2 \ B \ \frac{\partial^2 Q}{\partial t^2} = 3 B g S_o \ y_o^2 \left[R \ - \frac{\partial Q}{\partial w} \right] - \frac{2 g q_o}{C^2} \ \frac{\partial Q}{\partial t}.$$

We will be concerned only with the case for which R is not a function of w, and also we take $C = u_o / \sqrt{y_o \, S_o}$. We obtain finally

$$(B^2 y_o^3 \ g \ - q_o^2) \frac{\partial^2 Q}{\partial w^2} - 2 B q_o \ y_o \ \frac{\partial^2 Q}{\partial w \, \partial t} - y_o^2 B^2 \ \frac{\partial^2 Q}{\partial t^2}$$ (19a)

$$= 3 B^2 g S_o \ y_o^2 \left[\frac{\partial Q}{\partial w} - R \right] + \frac{2 B^3 y_o^3 g S_o}{q_o} \ \frac{\partial Q}{\partial t}$$

This is the final linearized equation of motion to be solved. It is solved for each channel segment in the network, with boundary conditions determined (if the magnitude of the channel segment is > 1) by flow in the two upstream channel segments. Flow at the upstream end of magnitude 1 segments is assumed to be identically zero. For the scenario discussed previously, the lateral inflow is:

$$r (w ,t) = r_o + v \, \delta(t),$$

where $\delta(\cdot)$ is a Dirac delta function. The initial condition for q is

$$q (w ,0) = q_o (w)$$

and denote the boundary condition by $q (0,t)$ for a particular channel segment, letting $w = 0$ correspond to the upstream end of that segment. Thus, we have

$$R (w ,t) = v \, \delta(t),$$ (19b)

$$Q (w ,0) = 0,$$ (19c)

and

$$Q (0,t) = q (0,t) - q_o (0).$$ (19d)

The equilibrium properties y_o and q_o in (19a) are functions of w, but we may expand these functions in Taylor series (with remainder) around the point $w = 0$ (upstream end of the channel segment) and again invoke the assumption that $\partial y_o / \partial w$ and $\partial q_o / \partial w$ are small compared to y_o and q_o. Again ignoring terms in Y, Q, $\partial y_o / \partial w$, and $\partial q_o / \partial w$ of order greater than first yields an equation identical to (19a) with $y_o (w)$ replaced by $y_o (0)$ and $q_o (w)$ replaced by $q_o (0)$. Henceforth, we will take q_o and y_o to stand for these functions evaluated at the upstream end of the corresponding channel segment.

The approach now is to restrict our attention solely to the deviation Q from equilibrium discharge. In particular, the IUH that we obtain will be in terms of deviations from equilibrium caused by our instantaneous pulse of rainfall. It is to be expected that the IUH so obtained will change with the values of the equilibrium conditions, which introduces non-linearity into the analysis. For further discussion of this problem, see Valdes, Fiallo, and Rodriguéz-Iturbe (1979) and Wang, Gupta, and Waymire (1981).

4.2. Solution of the General Flow Equation

We shall in the remainder of the paper be concerned with three cases of linear routing:
(i)   Translation routing -- all terms on the left-hand side of (19a) are neglected;

(ii)   Diffusion routing -- the second and third terms on the left-hand side of (19a) are neglected;

(iii) General linear routing -- all terms of (19a) are retained.

Solution techniques for the linear flow equation are discussed in detail by Harley (1967). We outline below the procedure for solving (19a) for case (iii) above with lateral inflow given by (19b) and initial and boundary conditions given by (19c) and (19d). Solutions for cases (i) and (ii) may be found in a similar manner; results are summarized in the next section.

The solution to (19a) is obtained by use of the Laplace transform. Let

$$Q^*(w,\theta) = \int\limits_0^\infty e^{-\theta t} Q(w,t)dt.$$

Then (19a) leads to

$$(B^2 y_o^3 g - q_o^2) \frac{\partial^2 Q^*}{\partial w^2}$$

$$- 2Bq_o y_o \theta \frac{\partial Q^*}{\partial w} - y_o^2 B^2 \theta^2 Q^*$$

$$= 3B^2 gS_o y_o^2 \left[\frac{\partial Q^*}{\partial w} - v \right] + \frac{2B^3 y_o^3 gS_o \theta}{q_o} Q^*,$$

or

$$(B^2 y_o^3 g - q_o^2)\frac{\partial^2 Q^*}{\partial w^2} - (2Bq_o y_o \theta + 3B^2 gS_o y_o^2)\frac{\partial Q^*}{\partial w}$$

$$- (y_o^2 B^2 \theta^2 + \frac{2B^3 y_o^3 gS_o \theta}{q_o})Q^* = -3B^2 gS_o y_o^2 v$$

The solution Q_c^* to the homogeneous problem under the requirement that $Q_c^*(w,\theta) \to 0$ as $w \to \infty$ is:

$$Q_c^*(w,\theta) = k(\theta)\exp[-\gamma(\theta;\beta)w]$$

where

$$\gamma(\theta;\beta) = (\beta_1\theta^2 + \beta_2\theta + \beta_3)^{1/2} - \beta_4\theta - \beta_3^{1/2}, \tag{20}$$

$$\beta_1 = [gy_o\,(1 - F_o{}^2)^2]^{-1},$$

$$\beta_2 = [u_o\,y_o\,(1 - F_o{}^2)^2]^{-1}\,S_o\,(F_o{}^2 + 2),$$

$$\beta_3 = [4y_o{}^2(1 - F_o{}^2)^2]^{-1}\,9S_o{}^2,$$

$$\beta_4 = [gy_o\,(1 - F_o{}^2)]^{-1}\,u_o\,,$$

$$F_o = u_o/\sqrt{gy_o}\,,$$

and where $k\,(\theta)$ is arbitrary. A particular solution is

$$Q_p^*(w\,,\theta) = \frac{3gS_o\,v}{\theta^2 + 2By_o\,gq_o{}^{-1}S_o\,\theta}$$

(constant with respect to w). The function $k\,(\theta)$ is determined by the boundary condition $Q^*(0,\theta)$. The final solution is

$$Q^*(w\,,\theta) = [Q^*(0,\theta) - Q_p^*(w\,,\theta)]\,\exp\,[-\gamma(\theta;\beta)w\,] + Q_p^*(w\,,\theta). \tag{21}$$

This may be inverted to obtain, for a channel segment of length x,

$$Q\,(x\,,t\,) = v\,\beta_o\,(t\,) + \int_0^t h\,(x\,,\tau\,;\,\beta)[Q\,(0,t-\tau) - v\,\beta_o\,(t-\tau)]d\,\tau \tag{22}$$

where

$$\beta_o\,(t\,) = 1.5u_o\,[1 - \exp(-2gS_o\,u_o{}^{-1}t\,)] \tag{23}$$

is related to Q_p^* by

$$Q_p^*(w\,,\theta) = \int_0^\infty e^{-\theta t}\,v\,\beta_o\,(t\,)dt \tag{24}$$

and h and γ are related by

$$\int_0^\infty e^{-\theta t}\,h\,(x\,,t\,;\beta)dt = e^{-\gamma(\theta;\beta)x} \tag{25}$$

The final form of h for general linear routing is given in Troutman and Karlinger (1985) and Harley (1967, equation 2.9).

4.3. Special Cases: Translation and Diffusion

The solution of (19a) for general linear routing (case (iii) in the previous section) is (22) with β_o given by (23), h is given by Harley (1967, Eqn. 2.9), and γ (related to h by (25)) is given by (20). For the two special cases, translation routing (case (i)) and diffusion routing (case (ii)), the solution also is given by (22), and again (24) and (25) hold, with β_o, h, and γ defined differently. For case (i),

$$\beta_o = 1.5u_o, \tag{26}$$

$$h(x,t;\beta) = \delta(t - \beta_1^{-1}x), \tag{27}$$

and

$$\gamma(\theta;\beta) = \beta_1^{-1}\theta, \tag{28}$$

where

$$\beta_1 = 1.5u_o. \tag{29}$$

For case (ii),

$$\beta_o = 1.5u_o, \tag{30}$$

$$h(x,t;\beta) = x(4\pi\beta_2 t^3)^{-1/2}\exp[-(4\beta_2 t)^{-1}(\beta_1 t - x)^2] \tag{31}$$

and

$$\gamma(\theta;\beta) = (2\beta_2)^{-1}[(\beta_1^2 + 4\beta_2\theta)^{1/2} - \beta_1], \tag{32}$$

where

$$\beta_1 = 1.5u_o \tag{33}$$

and

$$\beta_2 = (2S_o B)^{-1}q_o(1 - F_o^2). \tag{34}$$

We note for cases (ii) and (iii) that $\beta_o = 1.5u_o$ is constant (not a function of time), and that $\beta_o(t)$ for case (i) in (23) tends to $1.5u_o$ for large t. This constant value of $1.5u_o$ [also equal to β_1 in (29) and (33)] is an advective velocity (celerity), and β_2 in (34) is a diffusivity constant.

4.4. Flow Through Networks

Our goal is to obtain the conditional expectation of the network IUH, as shown in (5). However, because of problems associated with scaling the IUH to have area one (see Troutman and Karlinger, 1985, for details), we shall instead evaluate the conditional expectation of network response as determined by using the solution (22) for flow in individual links together with boundary conditions defined by the topological configuration of the network. This network response will depend on v, and will not in general have area one; $Q(t) = Q(t;T, X, B)$ henceforth will be used to denote this response and the dependence on v will not be shown explicitly. Upper case Q is used to denote both this network response and the discharge in an individual channel segment as in (22). It should be clear from the context which use is intended; the former depends only on time and the latter on distance and time. The desired IUH approximation then will be the conditional expectation of $Q(t)$ divided by the area under this curve, yielding a function with area one. Because of this scaling, this approximation will then not depend on v. It should be noted, however, that the approximation obtained this way will not be identical to that obtained by **first** scaling to have area one and **then** computing the conditional expectation.

5. LINEAR ROUTING AND THE WIDTH FUNCTION $L(x)$

When h is given by (27), we have the case of pure translation routing, and it would be expected that the IUH might be described entirely by basin geometry, with the hydraulic parameter $\beta_1 = 1.5u_o$ serving merely as a scaling factor. Under the assumptions of constant drainage density and constant celerity throughout the network, it is clear that the IUH at a given time t will be completely determined by the **number** of links some point of which lies at a channel distance x from the outlet, where $x = \beta_1 t$. Troutman and Karlinger (1984, 1985) call this number (as a function of x) the **width function** for the basin, although this label is not new (see for example Kirkby, 1976). The width $L(x;T,X)$ is defined more precisely as follows: With each point in a channel network we may associate a distance to the outlet of the basin, as measured longitudinally along the channel segments that water actually will follow in reaching the outlet. The width function is defined by proceeding in the opposite direction -- by first choosing a distance and then counting the number of links some point of which lies at this channel distance from the outlet. Thus, for any fixed distance $x > 0$, the width $L(x;T,X)$ is the number of links with the property that the distance of the downstream junction from the outlet is $\leq x$ and the distance of the upstream junction from the outlet is $> x$. The reasoning behind the label "width function" is as follows: If the basin's shape is approximately a rectangle, one dimension of which is the mainstream length, then the other dimension (or the "width" of the rectangle) should be roughly proportional to the average width function. The width function $L(x;T,X)$ for the simple hypothetical drainage basin in Figure 1a is illustrated in Figure 1b.

It turns out that the width function is closely related to discharge computed by **any** linear routing scheme, not just translation routing. Assume that flow through a given channel segment is governed by (22), that β_o is not a function of time (this is true for translation and diffusion, and almost true for large t under general linear routing), and that β_o and β are constant for all channel segments. Then the width function $L(x) \equiv L(x;T,X)$ and the basin discharge (at the outlet) $Q(t) \equiv Q(t;T,X,B)$ are related by

$$Q^*(\theta;\beta) = v\,\beta_o\,\theta^{-1}\gamma(\theta;\beta)L*(\gamma(\theta;\beta)), \qquad (35)$$

where

$$Q^*(\theta;\beta) = \int_0^\infty e^{-\theta t}\,Q(t)\,dt$$

and

$$L^*(\theta) = \int_0^\infty e^{-\theta x}\,L(x)\,dx.$$

Under translation routing [γ given by (28)], (35) yields

$$Q(t) = v\,\beta_o\,L(\beta_1 t); \qquad (36)$$

i.e., the shape of the IUH is identical to that of the width function, and distance is transformed to time by $x = \beta_1 t$.

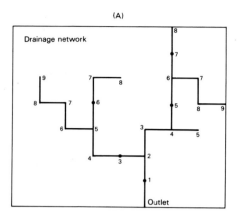

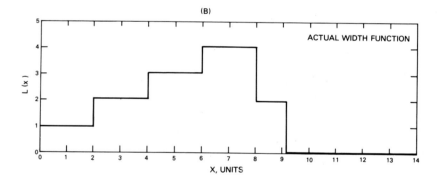

Figure 1. Hypothetical drainage basin and actual width function, $L(x)$.

This result in (35) is readily proven. For an individual channel segment, (21) is obeyed, with

$$Q_p^*(w,\theta) = v\,\beta_o\,\theta^{-1}.$$

Let $Q(x,t;\beta)$ be **total** flow at channel distance x from the outlet (i.e., $Q(x,t;\beta)$ is the sum of flows in all $L(x)$ channels), and let $Q^*(x,\theta;\beta)$ be the Laplace transform of this quantity. Then for any interval not containing a bifurcation or termination ($L(x)$ constant), we have for $\Delta x > 0$

$$Q^*(x,\theta;\beta) = v\,\beta_o\,\theta^{-1}[1 - e^{-\gamma(\theta;\beta)\Delta x}\,]L(x) + e^{-\gamma(\theta;\beta)\Delta x}\,Q^*(x+\Delta x,\theta;\beta).$$

For small Δx we have

$$Q^*(x,\theta;\beta) \doteq v\,\beta_o\,\theta^{-1}\gamma(\theta;\beta)\,\Delta x L(x) + (1 - \gamma(\theta;\beta)\Delta x)Q^*(x+\Delta x,\theta;\beta),$$

or

$$(\Delta x)^{-1}[Q^*(x+\Delta x,\theta;\beta) - Q^*(x,\theta;\beta)] \doteq \gamma(\theta;\beta)[Q^*(x+\Delta x,\theta;\beta) - v\,\beta_o\,\theta^{-1}L(x)].$$

Letting $\Delta x \to 0+$ gives

$$\frac{\partial Q^*(x,\theta;\beta)}{\partial x} = \gamma(\theta;\beta)[Q^*(x,\theta;\beta) - v\,\beta_o\,\theta^{-1}L(x)].$$

This ordinary differential equation is easily solved on each interval for which $L(x)$ is constant. Arbitrary constants are determined by matching solutions at points where $L(x)$ changes, and by requiring that $Q^*(x,\theta;\beta) \to 0$ as $x \to \infty$. The final solution, holding for all x, is

$$Q^*(x,\theta;\beta) = v\,\beta_o\,\theta^{-1}\,\gamma(\theta;\beta)\int_0^\infty e^{-\gamma(\theta;\beta)u}\,L(x+u)\,du.$$

At the outlet $x = 0$ so this expression gives (35).

6. APPLICATION OF METHODS IN BRANCHING THEORY TO NETWORK FLOW

Up to this point we have postulated a hydraulic routing model for flow in a channel segment of given length and with given hydraulic properties. If a number of segments were combined to form a network with a given topological configuration, flow at the outlet could be computed as a function of lateral inflow to the network. The problem is that the flow at the outlet is a rather complicated function of all the properties of the network; this functional dependence is depicted in equation (3). As discussed earlier, our strategy is to assume a probabilistic structure for (T,X,B) in (3), and to use this structure to derive simpler models for discharge. The primary purpose of this paper is to use branching theory to obtain the form of these simple models as a function of fundamental basin characteristics. In particular, branching theory is used to evaluate the conditional expectation in (5).

Branching theory is a widely researched area in stochastic processes. Excellent treatments may be found in books by Harris (1963) and Athreya and Ney (1972). A discrete time branching process is generated as follows: We begin with one individual at the zero'th generation, and at each stage, each individual in the population has probability p_k' of giving rise to *k* new individuals. It usually is assumed that all individuals of a given generation reproduce independently of each other. A continuous time branching process is similar, except it is assumed that lengths of individual lifetimes are continuous random variables.

The nature of the analogy between continuous time branching processes and random channel networks is straightforward. Time (with origin at the instant of birth of the individual in the zero'th generation) in the former is equivalent to channel distance (with origin at the downstream end of the outlet link) in the latter. In most of the theory of topologically random channel networks, it is assumed that no more than two links may come together at a single junction, which would be analogous to taking $p_k' = 0$, $k \neq 0,2$, in a branching process.

A branching process in which $p_o' > 0$, $p_2' > 0$, $p_o' + p_2' = 1$ is known as a **birth-death process** in the theory of stochastic processes. Thus, a bifurcation in a channel network corresponds to a birth and a termination (at the upstream end of an external link) corresponds to a death. The bifurcation probability, p is analogous to p_2' and the termination probability, q, is analogous to p_o'.

The term "birth" in branching theory may be given either of two interpretations: it may be assumed that the parent ceases to exist and gives rise to two new individuals, or that the parent continues to exist and gives birth to a single new individual. If the latter interpretation is made, another very useful analogy between branching processes and networks may be established: magnitude, N, of a channel network corresponds to cumulative population as time tends to infinity (i.e., to the total number of individuals that ever existed), provided the population becomes extinct at some finite time. Other quantities that are analogous are: (1) The width for a given channel distance x , $L(x)$, corresponds to population size at a given time; (2) mainstream length, $\tilde{D}$, corresponds to the time at which the population becomes extinct.

Generally speaking, there are two fundamental approaches to obtaining results in continuous time branching theory. In one approach, it is assumed that individual lifetimes (from birth to either death or splitting into two new individuals) are exponentially distributed and independent random variables. Population size (as a function of time) is then a **Markov process,** and a large body of theory exists for investigating such processes. In the second approach, the lifetime distribution is arbitrary, but the assumption that the lifetimes are independent and identically distributed is retained. The branching process is then called an **age-dependent** process, and the Markovian property is lost. A very powerful tool for investigating age-dependent branching processes is the theory of integral equations, much like the theory employed in examining renewal processes. For a general discussion of these two approaches, see Bailey (1964, chapters 6, 7, 8, and 16) and Karlin (1966, chapter 11).

7. THE WIDTH FUNCTION WITH EXPONENTIAL LINK LENGTHS

The theory of (Markov) birth-death processes leads to useful results for the width function with exponential link lengths and conditioning on fundamental basin characteristics $Z = N$, $Z = \tilde{D}$, and $Z = (N, \tilde{D})$ [(a), (d), and (e) in (6)]. In particular, we are interested in obtaining the first two moments of the width function, $L(x) = L(x;T,X)$, conditioned on Z. The relationship in (35) may be used to obtain the first two moments of $Q(t)$ from those of $L(x)$ under linear routing. For example, we have for the first moment

$$\int_0^\infty e^{-\theta t}\ \overline{Q}(t;z,\lambda)\ dt\ =\ v\,\beta_o\ \theta^{-1}\ \gamma(\theta;\beta)\int_0^\infty e^{-\gamma(\theta;\beta)x}\ \overline{L}(x;z,\lambda)dx$$

where $\overline{Q}$ is the mean of Q and $\overline{L}$ is the mean of L. The solution for $\overline{Q}$ here (properly normalized) is the IUH approximation we seek.

We first state precisely the following assumptions for this section: T, X_1, X_2, X_3, $\cdots$ are mutually independent; X_i are identically distributed as an exponential random variable with parameter η ($= mean^{-1}$); the network is drawn from a topologically random population with bifurcation probability $p = 1/2$. The vector Z is (a), (d), or (e) in (6).

Under these assumptions, the probabilistic properties of a channel network are identical to those of a tree generated by a birth-death process for which the birth rate is equal to the death rate (giving p = 1/2). A large number of results of interest may be obtained from the joint probability distribution of the width function $[L(x)]$, magnitude (N), and mainstream length $(\tilde{K})$. A generating function for these random variables is, for $d \geq x$,

$$\Psi(u,v;x,d) = \sum_{l=0}^{\infty} \sum_{n=0}^{\infty} u^l v^n \, P[L(x) = l, N = n, \tilde{D} \leq d] \tag{37}$$

$$= v \, \frac{v_1(v_2 - uv_3) + v_2(uv_3 - v_1) \, e^{-\eta\sqrt{1-v}\,(d-x)}}{v_2 - uv_3 + (uv_3 - v_1)e^{-\eta\sqrt{1-v}\,(d-x)}}$$

where

$$v_1 = 1 - \sqrt{1-v} \, , \, v_2 = 1 + \sqrt{1-v} \, , \, v_3 = v \, \frac{1-e^{-\eta\sqrt{1-v}\,(d-x)}}{v_2 - v_1 \, e^{-\eta\sqrt{1-v}\,(d-x)}}$$

This expression may be derived using the joint generating function of population size and cumulative population given in Bailey (1964, p. 125). Details will not be given here.

When $Z = N$, we obtain

$$E[L(x)|N = n] = q_n(x)/p_n$$

where $\{q_n(x)\}$ obeys

$$\sum_{n=0}^{\infty} q_n(x)v^n = \frac{\partial\Psi}{\partial u}\Big|_{u=1,d=\infty}$$

$$= (1 - \sqrt{1-v}) \, e^{-\eta\sqrt{1-v}\,x}$$

and p_n is given by (2); the variance of $(L(x)|N=n)$ may be found similarly. This generating function for $\{q_n(x)\}$ may be inverted to obtain an explicit expression for $E[L(x)|N=n]$; this expression is given by Troutman and Karlinger (1985), who obtained it using methods in age-dependent branching processes.

When $Z = \tilde{D}$, we have

$$E[L(x)|\tilde{D} = d] = \frac{\dfrac{\partial^2\Psi}{\partial u\,\partial d}\Big|_{v=1,u=1}}{\dfrac{\partial\Psi}{\partial d}\Big|_{v=1,u=1}} \tag{38}$$

$$= \begin{cases} 1 + \dfrac{\eta^2 x\,(d-x)}{2+\eta d}, & d \geq x \\ 0 & d < x \end{cases}$$

and, similarly, we obtain

$$Var\,[L\,(x\,)|\tilde{D}\,=d\,]= \begin{cases} \dfrac{\eta^2 x\,(d-x\,)}{2+\eta d} + \dfrac{\eta^4 x^2 (d-x\,)^2}{2(2+\eta d\,)^2}, d\,\geq x \\ 0,\quad d\,<x \end{cases} \tag{39}$$

It is interesting that the mean width function (38) is simply a parabola passing through $(0,1)$ and $(d,1)$ and peaking at $x\,=\dfrac{d}{2}$. Also note that the variance (3) is zero at $x\,=0$ and $x\,=d$, and peaks at $x\,=\dfrac{d}{2}$.

Finally, when $Z\,=(N,\tilde{D}\,)$, we obtain

$$E\,[L\,(x\,)|N\!=\!n\,,\tilde{D}\!=\!d\,] = r_n\,(x\,,d\,)/s_n\,(x\,,d\,) \tag{40}$$

where $\{r_n\,(x\,,d\,)\}$ and $\{s_n\,(x\,,d\,)\}$ obey

$$\sum_{n=0}^{\infty} r_n\,(x\,,d\,)\,v^{\,n} = \frac{\partial^2 \Psi}{\partial u\,\partial d}|_{x=1}$$

and

$$\sum_{n=0}^{\infty} s_n\,(x\,,d\,)v^{\,n} = \frac{\partial \Psi}{\partial d}|_{x=1'}$$

and similarly for the variance.

8. LINEAR FLOW WITH A GENERAL LINK LENGTH DISTRIBUTION

8.1. Method and Assumptions

We now employ the theory of age-dependent branching processes to obtain properties of the flow at the outlet of a network when the link length distribution is general.. This approach was utilized by Troutman and Karlinger (1985) to derive an integral-difference equation for the **expected** flow given fundamental basin characteristics; again, this expectation as a function of time serves as the IUH A generalization of the arguments they presented leads to an integral-difference equation for the **probability distribution** of flow under the assumption of translation routing, which may be used to obtain properties besides the expectation (such as the variance).

The theory of age-dependent processes is useful primarily for obtaining results when the vector of fundamental basin characteristics is one of the following: $Z\!=\!N,Z\!=\!(N,D\,)$, and $Z\!=\!(N,M\,)$. We will consider only those three cases in this section.

The following assumptions are made in this section: $T\,,X_1,\,X_2,\,X_3,\,\cdots$ are mutually independent; X_i is distributed with c.d.f. $F\,(\bullet;\alpha)$; the network is drawn from a topologically random population with p unspecified. The hydraulic properties of the links are assumed to be constant (β_o and β) throughout the network. Drainage density (lateral inflow) is constant throughout the basin. Routing through individual links is assumed to obey (22), with β_o and β constant. The vector Z is (a), (b), or (c) in (6).

Results in this section are readily generalized to the case for which internal and external links have different distribution and also different hydraulic parameters (see Troutman and Karlinger, 1984 and 1985, and Karlinger and Troutman, 1985). Note also that we have assumed that β_o is constant. This appears to rule out general linear routing, for which β_o was found to be a function of t. It is believed, however, that β_o in the general solution (equation (23)) may be set to its asymptotic (as $t \to \infty$) value of $3u_o/2$ with little error; this belief needs to be tested further. We shall in the remainder of the paper assume that β_o is constant for all cases.

8.2. Expected Flow with General Link Length Distribution

Troutman and Karlinger (1985) obtained the following integral-difference equation for $\overline{Q}_z(t;\alpha,\beta)$, the mean of $Q(t;T,X,\beta)$ given $Z=z$:

$$\overline{Q}_z(t;\alpha,\beta) = (1 - \int_0^t \overline{h}(\tau;\alpha,\beta)\,d\tau)v\,\beta_o + \sum_{z_1,z_2} P[Z_1 = z_1, Z_2 = z_2|Z=z] \qquad (41)$$

$$\int_0^t \overline{h}(t-\tau;\alpha,\beta)[\overline{Q}_{z_1}(\tau;\alpha,\beta) + \overline{Q}_{z_2}(\tau;\alpha,\beta)]d\,\tau,$$

where

$$\overline{h}(\tau;\alpha,\beta) = \int_0^\infty h(x,\tau;\beta)\,dF(x;\alpha)$$

is the impulse response averaged with respect to the link length distribution $F(\bullet;\alpha)$, and Z_1 and Z_2 are properties analogous to Z for the two subnetworks with outlets at the upstream end of the outlet link. The IUH approximation in (4) is then given by

$$\overline{Q}(t;z;\alpha,\beta) = \frac{\overline{Q}_z(t;\alpha,\beta)}{\int_0^\infty \overline{Q}_z(t;\alpha,\beta)dt}$$

A detailed derivation of (41) is presented in the Appendix of Troutman and Karlinger (1985).

For the fundamental basin characteristics listed in (6a,b,c) and under the assumption of a topologically random channel network, $P[Z_1 = z_1, Z_2 = z_2|Z=z]$ is symmetric in z_1 and z_2, and knowledge of z and z_1 determines the value of z_2 so (41) may be written as

$$\overline{Q}_z(t;\alpha,\beta) = 1 - \int_0^t \overline{h}(\tau;\alpha,\beta)d\,\tau \qquad (43)$$

$$+ 2 \sum_{z_1} P[Z_1 = z_1|Z = z] \int_0^t \overline{h}(t-\tau;\alpha,\beta)\,\overline{Q}_{z_1}(\tau;\alpha,\beta)d\,\tau$$

For the special case when $Z = N$, (43) becomes

$$\overline{Q}_n(t;\alpha,\beta) = 1 - \int_0^t \overline{h}(\tau;\alpha,\beta)d\,\tau$$

$$+ 2 \sum_{i=1}^{n-1} p_i\, p_{n-i}\, p_n^{-1} \int_0^t \bar{h}\,(t-\tau;\alpha,\beta)\ \bar{Q}_i\,(\tau;\alpha,\beta)d\,\tau,$$

where the magnitude of the link under consideration is n and the magnitudes of the two upstream links are i and $n-i$. We have used the fact that for topologically random networks

$$P\,[I=i\,|N=n\,] = p_i\, p_{n-i}\, p_n^{-1},$$

where I is the magnitude of one of the upstream links and p_j is given by (2).

Difference equations when $Z = (N,M)$ and $Z = (N,D)$ are more involved. Details may be found in Troutman and Karlinger (1984).

8.3. Distribution of Flow under Translation Routing

An argument similar to that used to obtain (41) (see Appendix of Troutman and Karlinger, 1985) may be employed to obtain a more general result for the width function. Define

$$\Phi_z\,(x\,,u\,;\alpha) = \sum_{k=0}^{\infty} u^k\ P\,[L\,(x\,;T\,,X) = k\,|Z=z\,]$$

to be the probability generating function (p.g.f.) of $L\,(x\,;T\,,X)$ conditional on $Z=z$. We have the following integral-difference equation:

$$\Phi_z\,(x\,,u\,;\alpha) = u\,[1 - F\,(x\,;\alpha)] + \sum_{z_1,z_2} P\,[Z_1 = z_1, Z_2 = z_2|Z = z\,] \qquad (44)$$

$$\int_0^x \Phi_{z_1}\,(x-\tau,u\,;\alpha)\ \Phi_{z_2}(x-\tau,u\,;\alpha)dF\,(\tau;\alpha)$$

The similarity between (41) and (44) is obvious. The product of Φ_{z_1}, Φ_{z_2} appearing in the integrand of (44) comes from the fact that the contributions from the left and right subnetworks are summed, and the p.g.f. of the sum of two independent random variables is the product of their p.g.f.'s. Differentiating (44) with respect to u and setting $u=1$ yields (41) for the width function.

Also, from (44) may be obtained an equation similar to (41) for higher order moments. For example, differentiating twice with respect to u and setting $u=1$ leads to

$$\bar{R}_z\,(x\,;\alpha) = 1 - F\,(x\,;\alpha) + \sum_{z_1,z_2} P\,[Z_1 = z_1, Z_2 = z_2|Z=z\,]$$

$$\int_0^x [\bar{R}_{z_1}(x-\tau;\alpha) + \bar{R}_{z_2}(x-\tau;\alpha) + 2\bar{L}_{z_1}(x-\tau;\alpha)\bar{L}_{z_2}(x-\tau;\alpha)]\,dF\,(\tau;\alpha)$$

where $\bar{R}_z\,(x\,;\alpha) = E\,[L^2(x\,;T\,,X)|Z=z\,]$ is the second moment of the width function and $\bar{L}_z$ is the first moment (defined earlier). Again, all results for the width function will apply to linear routing simply by using the transformation in (35).

8.4. Solution Techniques

It is seen in the previous two sections that age-dependent branching theory leads to integral-difference equations that must be solved, usually numerically, to obtain the quantities of interest. There are special cases for which it is possible to obtain an explicit solution to (41) and (44). Troutman and Karlinger (1985) give an explicit solution for $\overline{Q}_z(t;\alpha,\beta)$ in (41) when the link lengths are gamma distributed with an integer-valued shape parameter. Included in this class of distributions is the exponential distribution. Analogous explicit solutions for (44) may be obtained.

A very powerful tool for solving the integral-difference equations is the use of transforms (or generating functions) (Troutman and Karlinger, 1984, 1985). Define first the following Laplace transforms:

$$\overline{h}^*(\theta;\alpha,\beta) = \int_0^\infty e^{-\theta t}\ \overline{h}(t;\alpha,\beta)dt$$

and

$$\overline{Q}_z^*(\theta;\alpha,\beta) = \int_0^\infty e^{-\theta t}\ \overline{Q}_z(t;\alpha,\beta)dt.$$

From (42) we see that $\overline{h}^*$ may be obtained from

$$\overline{h}^*(\theta;\alpha,\beta) = F^*(\gamma(\theta;\beta);\alpha),$$

where F^* is the Laplace transform of the link length distribution,

$$F^*(\theta;\alpha) = \int_0^\infty e^{-\theta x}\ dF(x;\alpha),$$

and $\gamma(\theta;\beta)$ for the three linear routing schemes mentioned before are given in (20), (28), and (32).

Taking the Laplace transform of (43) gives

$$\overline{Q}_z^*(\theta;\alpha,\beta) = \theta^{-1}[1 - \overline{h}^*(\theta;\alpha,\beta)] \tag{45}$$

$$+ 2\overline{h}^*(\theta;\alpha,\beta) \sum_{z_1} P[Z_1 = z_1|Z = z]\overline{Q}_{z_1}^*(\theta;\alpha,\beta)$$

using the convolution theorem; the initial condition is

$$\overline{Q}_1^*(\theta;\alpha,\beta) = \theta^{-1}[1 - \overline{h}^*(\theta;\alpha,\beta)].$$

Equation (45) is extremely useful for numerical evaluation of the IUH because the convolution integral has been reduced to a multiplication by the Laplace transform. This difference equation is again solved recursively for a set of θ's, and numerical inversion of the transform gives the desired IUH

When Z is N, (45) becomes

$$\overline{Q}_n^*(\theta;\alpha,\beta) = \theta^{-1}[1 - \overline{h}^*(\theta;\alpha,\beta)] \tag{46}$$

$$+ 2\overline{h}^{*}(\theta;\alpha,\beta) \sum_{i=1}^{n-1} p_{n}^{-1}\ p_{i}\ p_{n-i}\ \overline{Q}_{i}^{*}(\theta,\alpha,\beta).$$

We may go even further in solving (46) by transforming with respect to n. Put

$$\overline{Q}^{**}(s,\theta;\alpha,\beta) = \sum_{n=0}^{\infty} s^{n}\ 4^{-n}\ p_{n}\ \overline{Q}_{n}^{*}(\theta;\alpha,\beta).$$

Multiplying (46) by $s^{n}\ 4^{-n}\ p_{n}$ and summing gives

$$\overline{Q}^{**}(s,\theta;\alpha,\beta) = \frac{\overline{p}(s)[1 - \overline{h}^{*}(\theta;\alpha,\beta)]}{\theta[1 - 2h^{*}(\theta;\alpha,\beta)\ p(s)]} \tag{47}$$

Analogous expressions for the cases $Z = (N,D)$ and $Z = (N,M)$ may be obtained (Troutman and Karlinger, 1984).

We also may express (44) in a more convenient form using generating functions. Assume again that $Z = N$, and set

$$\Phi^{**}(s,x,u;\alpha) = \sum_{n=0}^{\infty} s^{n}\ 4^{-n}\ p_{n}\ \Phi_{n}(x,u;\alpha).$$

Then, (44) leads to

$$\Phi^{**}(s,x,u;\alpha) = u\overline{p}(s)[1 - F(x;\alpha)] + \int_{0}^{x} [\Phi^{**}(s,x-\tau,u;\alpha)]^{2}\ dF(\tau;\alpha);$$

the presence of the squared term in the integrand, however, makes this equation impossible to solve explicitly.

The transform for $Z = N$ in (47) and the analogous transforms for $Z = (N,M)$ and $Z = (N,D)$ in most cases do not lend themselves to explicit inversion, but they are very useful for examining the behavior of moments of the IUH. Setting the transform argument θ in (45) equal to zero gives a difference equation for the area (zero-th moment) of $\overline{Q}_{z}\ t;\alpha,\beta)$. Taking derivatives with respect to θ and then setting $\theta = 0$ will yield difference equations for higher order moments of $\overline{Q}_{z}(t;\alpha,\beta)$. It is interesting that the r-th moment of $\overline{Q}_{z}(t;\alpha,\beta)$ is a function of the first $r+1$ moments of $\overline{h}(t;\alpha,\beta)$.

9. ASYMPTOTIC RESULTS

9.1. General Considerations

Many of the results in this paper are presented in terms of generating functions or Laplace transforms, inversion of which does not yield simple closed form solutions. One nice property of the transforms, however, is that they often provide a convenient means for examining asymptotic behavior of the solutions, and the form of the asymptotic results usually turns out to be much simpler than that of the nonasymptotic results. The term "asymptotic" will be taken to have one of several meanings here; we will let one or more of the following three variables tend to infinity; N (number of first-order streams), D (topological diameter), or $\tilde{D}$ (mainstream length).

9.2. Expected Width Function with Exponential Links

We first consider asymptotic behavior of the width function with exponential link lengths. The asymptotic form as $n \to \infty$ of $E[L(x)|N=n]$, scaled to have area (integral with respect to x) one, is a Weibull probability density function. (This form also holds for nonexponential link lengths and may be derived using age-dependent branching theory. This will be considered in more detail shortly.) The form of $E[L(x)|N=n,\tilde{D}=d]$ as both n and $d \to \infty$ may be derived from (37). We first present the following intermediate result, which is of interest in itself:

$$\lim_{d \to \infty} E\left[exp\left(-ud^{-1}L(cd)-vd^{-2}N\right)|\tilde{D}=d\right]$$

$$= \lim_{d \to \infty} \frac{\frac{\partial \Psi}{\partial d}(u',v';x',d)\big|_{u'=exp(-ud^{-1},v'=exp(-vd^{-2},x'=cd)}}{\frac{\partial \Psi}{\partial d}(u',v';x',d)\big|_{u'=1,v'=1,x'=dc}}$$

$$= A/B, 0 \le c \le 1$$

where

$$A = \eta^2 v \{uc(u+2\sqrt{v})exp(-\eta c\sqrt{v}) + (4v-2cu^2)exp(-\eta\sqrt{v}) + uc(u-2\sqrt{v})exp(-\eta(2-c)\sqrt{v})\}$$

$$B = \{(2\sqrt{v}+u-u)exp(-\eta(1-c)\sqrt{v}) - u\,exp(-\eta c\sqrt{v}) + (u-2\sqrt{v})exp(-\eta\sqrt{v})\}^2.$$

This is essentially the asymptotic joint Laplace transform of the width function and the magnitude, given mainstream length; c represents the ratio x/d. The result leads to, for large d and n,

$$E[L(x)|N=n,\tilde{D}=d]$$

$$\frac{2d}{\eta}D_y\left[y^{-1/2}\sum_{i=2}^{\infty}\{(i-1)(i-c)exp\left[-\frac{\eta^2}{4y}(i-1)^2\right] + (i-c)(i+c-2)\right.$$

$$exp\left[-\frac{\eta^2}{4y}(i+c-2)^2\right] - 2i^2exp\left(-\frac{\eta^2}{4y}i^2\right)\}]$$

$$\div D_y\{y^{-3/2}\sum_{i=1}^{\infty}i^2\,exp\left[-\frac{\eta^2}{4y}i^2\right]\}\big|_{y=\frac{n}{d^2},c=\frac{x}{d}}$$

here $D_y = \dfrac{d}{dy}$.

9.3. Expected Flow with General Links

Asymptotic results for expected discharge with general link length distribution are discussed in detail in Troutman and Karlinger (1984, 1985). The most interesting result in those papers is that the expected IUH as a function of time, conditional on $Z=N$, tends to a Weibull probability density function as N gets large. First define

$$\alpha^* = \int_0^\infty x \, dF(x;\alpha), \tag{48}$$

the mean of the link length distribution, and define

$$\beta^* = [\gamma'(0,\beta)]^{-1}, \tag{49}$$

where γ' denotes the partial derivative of $\gamma(\theta,\beta)$ with respect to θ. Note that α^* is a function of α and β^* is a function of hydraulic parameters β, and both α^* and β^* are scalars. We have for the three linear routing methods:

$$\beta^* = 1.5u_o \tag{50}$$

Also, let $\lambda^* = \beta^*/\alpha^*$. The asymptotic expression for the IUH for $Z = N$, is:

$$\overline{Q}(t;n;\alpha,\beta)(2n)^{-1}(\lambda^{*2}t)\exp[-(4n)^{-1}(\lambda^* t)^2], t \geq 0. \tag{51}$$

The time to peak (mode) for $\overline{Q}(t;n;\alpha,\beta)$ is asymptotically

$$t_p = \sqrt{2n} \ \lambda^{*-1} \tag{52}$$

and thus, the peak is

$$\overline{Q}(t_p;n;\alpha,\beta) = (2ne)^{-1/2}\lambda^*.$$

Asymptotic results for $Z = (N,D)$ and $Z = (N,M)$ also are presented in Troutman and Karlinger (1984, 1985); however, the manner in which the limits were evaluated makes it uncertain whether these results will be as useful as the previous result for $Z = N$. For all three cases the following conclusions, which are hydrologically very revealing, hold (Troutman and Karlinger, 1985):

(1) All pertinent hydraulic information is contained in the single parameter β^*, given by (50) for the three linear routing schemes we have considered. For all three forms of routing, β^* represents celerity.

(2) The only property of the link length distribution that is important is the mean, α^*. No assumptions were made concerning distribution.

(3) The effects of α^*, β^*, and n enter only as a single scaling parameter in all three of the situations $Z = N, Z = (N,D)$ and $Z = (N,M)$; that parameter is $\lambda^{*-1}\sqrt{n}$ in the first case and $\lambda^{*-1}n$ in the remaining cases; where $\lambda^* = \beta^*/\alpha^*$. The parameter λ^{*-1} actually represents a mean holding time per link, in the sense that α^* is mean link length and β^* is an average celerity.

(4) All linear routing schemes lead to the same asymptotic IUH, suggesting that one does not gain much by using more complex routing techniques. Only a factor that measures average celerity, β^*, is important.

10. SIMULATION STUDY

10.1. Framework

The primary purpose of the simulation study in this section is (1) to look at how well the IUH's based on fundamental basin characteristics, as computed using methods in this paper, approximate the actual IUH, and (2) to compare different fundamental basin characteristics. Previous studies (Troutman and Karlinger, 1985; and Karlinger and Troutman, 1985) indicate that agreement between actual and approximate is good, but evaluations are based on examples, and no systematic examinations have been presented.

Some of the analytic results on variance of the width function presented in this paper are useful in comparing methods and determining uncertainty, but these results are of limited value for two reasons. First, we are interested in uncertainty in the IUH, and the IUH is computed by normalizing to obtain area one. This normalizing involves division by a quantity which is itself random (Troutman and Karlinger, 1985). Thus, we are really interested in the variance of the ratio of two random variables, and analytic results give the variance of only the numerator.

Second, the analytic results give pointwise variance as a function of distance or time. For hydrologic applications we are usually more interested in uncertainty in various functionals of the hydrograph, such as the peak and time to peak. Analytic results on these quantities are not yet available.

In this study we will be concerned with four quantities. Let $I(t)$ stand for an actual IUH (scaled to have area one). Define the peak

$$P = \sup_{\{0 \,\leq\, t \,<\, \infty\}} \{I(t)\}, \tag{53}$$

time to peak (mode)

$$T = \inf_{0 \,\leq\, t \,<\, \infty} \{t : I(t) = P\}, \tag{54}$$

mean

$$M = \int_0^{\infty} t I(t)\,dt, \tag{55}$$

and standard deviation

$$S = [\int_0^{\infty} (t - M)^2 I(t)\,dt]^{1/2}. \tag{56}$$

There are many other properties of hydrographs that might be of interest hydrologically, and in fact research is needed to identify which properties are important. Hydrologists typically emphasize P and T; M is a measure of central tendency, S is a measure of dispersion. We shall use P, T, M, and S to signify properties of the actual IUH and $\hat{P}$, $\hat{T}$, $\hat{M}$, and $\hat{S}$ to denote corresponding values obtained from the IUH approximation. Next, let

$$b_K = E(K - \hat{K})$$
$$v_K = E[(\hat{K} - E(\hat{K}))^2]$$

and

$$r_K = E[(\hat{K} - K)^2] = v_K + b_K^2,$$

where K is P, T, M, or S. We see that r_K is mean-square error, v_K is variance, and b_K is bias.

We shall assume here that the actual IUH is generated by translation routing with $\beta_1 = 1.0$ (known) in (22), and that the basin comes from a topologically random population with $N = 5$ sources and with link lengths that are independent and exponentially distributed with mean 1.0. Approximate IUH's are computed based on the following sets of fundamental basin characteristics Z^*:

(a) $N, \overline{X}, \beta_1$

(b) $N, D, \overline{X}, \beta_1$

(c) $N, M, \overline{X}, \beta_1$

(d) $\tilde{D}, \overline{X}, \beta_1$

(e) $N, \tilde{D}, \overline{X}, \beta_1$ and

(f) $M, R_A, R_B, R_L, \overline{L}, V = \beta_1$.

Here we let $\overline{X}$ = sample mean link length and $\overline{L}$ is sample mean first-order link length. The IUH approximation for (a) through (c) is computed using (41), that for (d) using (38), that for (e) using (40), and that for (f) is computed by the methods in Rodriguéz-Iturbe and Valdés (1979). The area ratio, R_A, for method (f) is computed by assuming that area draining into any particular segment of channel is proportional to the length of that segment (constant drainage density). Also, because we are considering only basins with $N = 5$ sources, results for methods (b) and (c) will be identical. This is due to the fact that when $N = 5$, $D = 5$ if and only if $M = 2$, and $D = 4$ if and only if $M = 3$.

10.2. Results

One hundred basins were generated at random by first selecting a topological configuration at random (there are 14 equally likely configurations for basins with $N = 5$ sources) and then selecting the link lengths independently from an exponential distribution (there are nine links in basins with five sources). The actual IUH was obtained and K computed, where K is P, T, M, and S. Approximate IUH's were obtained and $\hat{K}$ computed for each of the six sets of characteristics (a) through (f). The quantities b_K, v_K, and r_K then were estimated by (trimmed) sample means for the 100 basins. Five-percent-trimmed means (sample mean computed after deleting the 5 percent highest and 5 percent lowest values; see Rosenberger and Gasko, 1983) are presented in Table 1.

Trimmed means are employed here because untrimmed estimates of b_K, v_K, and r_K were being overly influenced by a few outstanding values. Table 2 summarizes the methods that performed best with respect to the different criteria (bias, variance, and mean-squared error).

Plots of actual and approximate IUH's for two realizations of the simulation are shown in Figure 2. These two realizations represent somewhat extreme cases. Figure 2(a) corresponds to a basin with one rather long external link, and Figure 2(b) corresponds to a basin with a low mean link length.

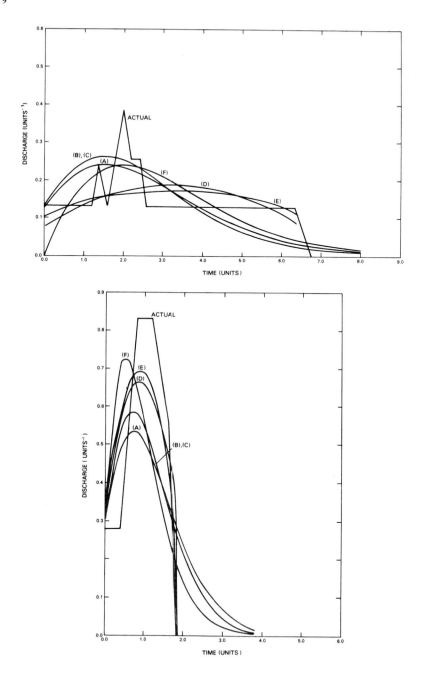

Figure 2. Actual and approximate IUH's for two realizations of simulation.

10.3. Discussion

Several tendencies immediately become apparent in Table 1. First, all methods tend to underpredict the peak (positive bias). The reason for this, at least for methods (a) through (e) developed in this paper, is obvious. These methods are based on conditional expectations, and the expectation of a stochastic process will not exhibit the fluctuations of the process itself; hence, the peak of the random process usually will be greater. In our analysis we have used $\sup_t E\left[Q(t)|Z=z\right]$ for peak prediction, but ideally we should use $E\left[\sup_t Q(t)|Z=z\right]$; analytical results for the latter seem to be more difficult to obtain.

TABLE 1. Estimates of Bias, Variance, and Mean-Squared Error Obtained from Simulation Study

K=	P	T	M	S
		Mean of Actual		
	0.398	2.160	2.692	1.314
		bias (b_K)		
(a)	0.141	0.342	0.041	-0.354
(b),(c)	.143	.324	.027	-.360
(d)	.129	-.469	.090	.091
(e)	.123	-.470	.113	-.037
(f)	.112	.030	-.356	-.365
		variance (v_K)		
(a)	0.008	1.435	0.341	0.103
(b),(c)	.007	1.463	.313	.089
(d)	.004	1.475	.089	.019
(e)	.004	1.485	.052	.007
(f)	.020	1.538	.396	.115
		Mean-squared error (r_K)		
(a)	0.028	1.552	0.343	0.228
(b),(c)	.027	1.568	.313	.219
(d)	.021	1.695	.097	.028
(e)	.020	1.706	.064	.009
(f)	.033	1.539	.523	.248

TABLE 2. Summary of Methods that Performed Best with Respect to Different Criteria in Simulation Study

K =	P	T	M	S
Bias	(f)	(f)	(b),(c)	(e)
Variance	(d),(e)	(a)	(e)	(e)
Overall	(e)	(f)	(e)	(e)

Methods (a), (b), and (c) tend to underpredict the time to peak, while the opposite is true of (d) and (e); (f) is nearly unbiased in predicting T. For method (d), $\hat{T}$ is always $\tilde{d}/2 =$ (mainstream length)/2, and for method (e) $\hat{T}$ usually is close to $\tilde{d}/2$. The fact that $\hat{T}$ for these methods tends to overpredict T is consistent with the observation that actual hydrographs often are skewed to the right, so that the mode generally is shifted to the left.

Bias in predicting M is small for methods (a) through (e) and negative for method (f). Bias in predicting S is small for (d) and (e) and negative for the remaining methods.

The variances in predicting P and T generally are comparable for all the methods, but they are somewhat higher for method (f). For predicting M and S, methods (d) and (e) greatly outperform the other methods with respect to variance.

Method (e) has the smallest mean-squared error and method (d) the next smallest for predicting P, M, and S. These two methods are by far the best for predicting M and S. Method (f) has the smallest mean-squared error for predicting T.

Overall, method (e) seems to stand out the most in predicting P, M, and S, but the severe bias in predicting T is disturbing. Method (f) exhibits the smallest bias in predicting T, and method (a) the smallest variance. Generally, there is not much difference in the performance of (a) and that of (b) and (c), so that the additional complexity in using the latter may make (a) more desirable.

11. SUMMARY AND CONCLUSIONS

1. Methods in branching theory provide a powerful theoretical tool for averaging drainage-basin properties to obtain IUH approximations under linear channel routing.

2. Analytically tractable solutions for the conditional expectations of discharge given Z are available; Z consists of one or two of the following fundamental basin characteristics: N (number of first-order streams), D (topological diameter), M (order), and $\tilde{D}$ (mainstream length). This conditional expectation serves as the IUH approximation.

3. Under translation routing, the shape of the IUH is identical to that of the width function for the basin, where for given channel distance x, the width function is the number of links some point of which lies at distance x from the outlet.

4. Asymptotic considerations indicate that for all linear routing schemes considered and for given $Z=N$, $Z=(N,D)$ and $Z=(N,M)$, the following conclusions hold for the expected IUH: All pertinent hydraulic information is contained in a single scalar parameter (β^*) that represents celerity, the only important property of the link length distribution is the mean (α^*), and the effects of α^*, β^*, and N enter only as a single scaling parameter.

5. All linear routing schemes, again for $Z=N$, $Z=(N,D)$, and $Z=(N,M)$, lead to the same asymptotic expected IUH, which suggests that translation routing may be entirely adequate for routing through large networks. This point establishes the value of the width function in hydrologic applications.

6. The asymptotic expected IUH for given $Z=N$ is a Weibull probability density function, with time to peak given by $\sqrt{2N}\ \alpha^*/\beta^*$.

7.   A simulation study indicates that all IUH approximation methods considered in the
     paper tend to underpredict the peak flow, P. This study also indicates that the IUH
     based on $(N,\hat{D})$ generally performs better than the other methods in predicting P, M,
     and S, but that this IUH exhibits a severe bias in predicting T. The IUH approxima-
     tion developed by Rodriguéz-Iturbe and Valdés (1979) has the smallest bias in predicting
     T, and the IUH based on N has the smallest variance.

ACKNOWLEDGMENTS

The authors appreciate the careful reviews and helpful comments of A.V. Vecchia and R.L.
Naff.

REFERENCES

Athreya, K.B. and Ney, P.E., 1972. *Branching Processes*, Springer-Verlag, Berlin.

Bailey, N.T.J., 1964. *The Elements of Stochastic Processes with Applications to the Natural
Sciences*, John Wiley and Sons, NY.

Dooge, J.C.I. and Harley, B.M., 1967. Linear Routing in Uniform Open Channels, *Proc. Int'l.
Hydrol. Symp.*, v. 1, pp. 8.1-8.7.

Gupta, V.K., Waymire, E., and Wang, C.T., 1980. A Representation of an Instantaneous
Unit Hydrograph From Geomorphology, *Water Resources Research*, 16(5), pp. 855-862.

Gupta, V.K., and Waymire, E., 1983. On the Formulation of an Analytical Approach to
Hydrologic Response and Similarity at the Basin Scale, *J. Hydrology*, 65, pp. 95-124.

Harley, B.M., 1967. Linear Routing in Uniform Open Channels, M.S. Thesis, National
University of Ireland.

Harris, T.E., 1963. *The Theory of Branching Processes*, Springer-Verlag, Berlin.

Karlin, S., 1966. *A First Course in Stochastic Processes*, Academic Press, NY.

Karlinger, M.R., and Troutman, B.M., 1985. An Assessment of the Instantaneous·Unit
Hydrograph Derived from the Theory of Topologically Random Networks, *Water Resources
Research*, 21(11), pp. 1693-1702.

Keefer, T.N., and McQuivey, K.S., 1974. Multiple Linearization Flow Routing Model, *J.
Hydraul. Div., ASCE*, 100(HY7), pp. 1031-1046.

Kirkby, M.J., 1976. Tests of the Random Network Model and Its Applications to Basin
Hydrology, *Earth Surf. Proc.*, 1, pp. 197-212.

Kirshen, D.M., and Bras, R.L., 1983. The Linear Channel and its Effect on the Geomorphic
IUH, *J. Hydrology*, 65, pp. 175-208.

Leopold, L.B., and Maddock, T., Jr., 1953. *The Hydraulic Geometry of Stream Channels and Some Geomorphologic Implications*, U.S. Geol. Survey Prof. Paper 252, 56 pp.

Lienhard, J.H., 1964. A Statistical Mechanical Prediction of the Dimensionless Unit Hydrograph, *J. Geophys. Res.*, 69(24), pp. 5231-5238.

Pilgrim, P.H., 1977. Isochrones of Travel Time and Distribution of Flood Storage from a Tracer Study on a Small Watershed, *Water Resources Research*, 13(3), pp. 587-595.

Rodríguez-Iturbe, I., and Valdés, J.B., 1979. The Geomorphologic Structure of Hydrologic Response, *Water Resources Research*, 15(6), pp. 1409-1420.

Rosenberger, J.L., and Gasko, M., 1983. Comparing Location Estimators: Trimmed Means, Medians, and Trimean, in *Understanding Robust and Exploratory Data Analysis*, edited by D.C. Hoaglin, F. Mosteller, and J.W. Tukey, John Wiley & Sons, NY.

Shreve, R.L., 1966. Statistical Law of Stream Numbers, *J. Geology*, 74, pp. 17-37.

Shreve, R.L., 1967. Infinite Topologically Random Channel Networks, *J. Geology*, 75, pp. 178-186.

Shreve, R.L., 1969. Stream Lengths and Basin Areas in Topologically Random Channel Networks, *J. Geology*, 77, pp. 397-414.

Shreve, R.L., 1974. Variation of Mainstream Length with Basin Area in River Networks, *Water Resources Research*, 10, pp. 1167-1177.

Smart, J.S., 1968. Statistical Properties of Stream Lengths, *Water Resources Research*, 4, pp. 1001-1014.

Smart, J.S., 1972. Channel Networks, in *Advances in Hydroscience*, edited by Ven Te Chow, Academic Press, NY.

Smart, J.S., 1978. Analysis of Drainage Network Composition, *Earth Surf. Proc.*, 3, pp. 129-170.

Smart, J.S., and Werner, C., 1976. Analysis of the Random Model of Drainage Basin Composition, *Earth Surf. Proc.*, pp. 219-233.

Troutman, B.M., and Karlinger, M.R., 1984. On the Expected Width Function for Topologically Random Channel Networks, *J. of Applied Probability*, 21, pp. 836-849.

Troutman, B.M., and Karlinger, M.R., 1985. Unit Hydrograph Approximations Assuming Linear Flow Through Topologically Random Channel Networks, *Water Resources Research*, 21(5), pp. 743-754.

Valdés, J.B., Fiallo, Y., and Rodríguez-Iturbe, I., 1979. A Rainfall-Runoff Analysis of the Geomorphologic IUH, *Water Resources Research*, 15(6), pp. 1421-1434.

Wang, C.T., Gupta, V.K., and Waymire, E., 1981. A Geomorphic Synthesis of Nonlinearity in Surface Runoff, *Water Resources Research*, 17(3), pp. 545-554.

Werner, C., and Smart, J.S., 1973. Some New Methods of Topologic Classification of Channel Networks, *Geographical Analysis*, 5, pp. 271-295.

10 INCORPORATION OF CHANNEL LOSSES IN THE GEOMORPHOLOGIC IUH

Mario Diaz-Granados
Rafael L. Bras
Juan B. Valdés

ABSTRACT

The infiltration losses along the stream channels of a basin are included into the Instantaneous Unit Hydrograph (IUH). The IUH is derived as a function of the basin geomorphological and physiographic characteristics, and the response of the individual channels to upstream and lateral inflows. This response is obtained by solving the linearized continuity and momentum equations, including approximate infiltration losses terms, for the boundary conditions established by the definition of a linear system response to an instantaneous unit input. A methodology is proposed for the estimation of the parameters involved in the channel response. Based on this result, a procedure is suggested to include infiltration losses in the common linear reservoir representation of channel segments. Comparisons indicate that this approximation is adequate under certain conditions.

For the first time channel infiltration is explicitly included in an analytical physically based linear model of channel response potentially useful in traditional hydraulic routing problems.

1. INTRODUCTION

Recently, methodologies have been proposed to relate river response to basin geomorphology (Rodriguéz-Iturbe and Valdés, 1979), which are useful in the estimation of the hydrologic behavior in regions with sparse or no data. The Instantaneous Unit Hydrograph, IUH, is interpreted as the probability density function (PDF) of the travel time spent by a drop to reach the outlet of the basin, which is a function of the geomorphology quantified by the Horton numbers, and the response of individual channels, assumed to behave like linear reservoirs. This IUH is called the Geomorphologic IUH (GIUH).

In its derivation, the Strahler's channel ordering scheme is used, which allows us to express

V. K. Gupta et al. (eds.), Scale Problems in Hydrology, 217–243.
© *1986 by D. Reidel Publishing Company.*

the cumulative density function (CDF) of the time that a drop takes to travel to the outlet of

$$P(T_B \leq t) = \sum_{s \in S} P(T_s \leq t)P(s)$$

where $P(\cdot)$ represents the probability of the event given in parenthesis, T_B is the travel time to the outlet of the basin, T_s is the travel time through a path s, belonging to S, the set of all possible paths that a drop, falling randomly on the basin, may follow to reach the outlet. The travel time, T_s, in a particular path, must be equal to the sum of travel times in the elements of the path:

$$T_s = T_{r(i)} + \cdots T_{r(\Omega)}$$

where $T_{r(i)}$ is the travel time in a stream of order i, Ω is the order of the basin.

Given that there exists several streams of a given order, $T_{r(i)}$ may be considered an independent random variable with a given probability density function, $f_T{}^{(i)}(t)$, so that the cumulative density function of T_s is the convolution of the individual cumulative density functions, $F_T{}^{(i)}(t)$:

$$F_T^s(t) = F_T{}^{(i)}(t) * \cdots * F_T{}^{(\Omega)}(t)$$

where * indicates the convolution operation. The probability of a given path s is:

$$P(s) = \Theta_i \cdot P_{ij} \cdots P_{k\Omega}$$

where Θ_i is the probability that a drop falls in an area draining to a stream of order i and P_{ij} is the transition probability from streams of order i to streams of order j. Rodriguéz-Iturbe and Valdés (1979) show that the initial and transition probabilities are functions only of the geomorphology of the basin, namely the bifurcation and area ratios. Table 1 summarizes these probabilities for a basin of order 3. The GIUH is, then:

$$h(t) = \frac{dP(T_B \leq t)}{dt} \tag{1}$$

$$= \sum_{s \in S} f_T{}^{(i)}(t) * \cdots * f_T{}^{(\Omega)}(t) P(s)$$

Rodriguéz-Iturbe and Valdés (1979) argue for an exponential behavior of the individual channels:

$$f_T{}^{(i)}(t) = \lambda_i \; e^{-\lambda_i t} \tag{2}$$

where

$$\lambda_i = v / \bar{L}_i$$

and the assumption that for a given rainfall-runoff event the velocity, v, at any moment is approximately the same throughout the whole drainage network, is made (Pilgrim, 1977).

Kirshen and Bras (1981) suggested that a physically based form of $f_T^i(t)$ can be obtained by

TABLE 1. Probabilities for a Basin of Order 3

$$\Theta_1 = \frac{R_B^2}{R_A^2}$$

$$\Theta_2 = \frac{R_B}{R_A} - \frac{R_B^3 + 2R_B^2 - 2R_B}{R_A^2(2R_B - 1)}$$

$$\Theta_3 = 1 - \frac{R_B^3}{R_A} - \frac{R_B^3 - 3R_B^2 + 2R_B}{R_A^2(2R_B - 1)}$$

$$P_{12} = \frac{R_B^2 + 2R_B^2 - 2}{2R_B^2 - R_B}$$

$$P_{13} = \frac{R_B^2 - 3R_B + 2}{2R_B^2 - R_B}$$

$$P_{23} = 1$$

linearization of the equations of motion in a channel. In this paper, an approximate way to include channel infiltration losses is given following the approach taken by the above authors. Such losses can be important in semi-arid regions of the world.

2. LINEAR SOLUTION TO THE EQUATIONS OF MOTION

The one-dimensional equations of motion for unsteady flow in a wide rectangular open channel including infiltration losses are given by:

Continuity:

$$\frac{\partial q}{\partial x} + \frac{\partial y}{\partial t} = - q_I(x,t) \tag{3}$$

Momentum:

$$\frac{\partial y}{\partial x} + \frac{v}{g}\frac{\partial v}{\partial x} + \frac{1}{g}\frac{\partial v}{\partial t} - \frac{v}{gy} q_I(x,t) = S_o - S_f \tag{4}$$

where

g   = gravitational acceleration $[\, LT^{-2}\,]$

v   = mean velocity $[\, LT^{-1}\,]$

y   = depth $[\mathrm{L}]$

q $=$ discharge per unit width $[\, L^2 T^{-2}] \equiv vy$

S_o $=$ slope of the channel bottom

S_f $=$ friction slope

x $=$ space coordinate, measured downstream along the channel $[L]$

t $=$ time coordinate $[T]$

$q_I(x,t)$ $=$ infiltration rate $[\, LT^{-1}\,]$

The Chezy formula is used to describe the frictional effects,

$$S_f \;\sim\; \frac{v^2}{C^2 y} \tag{5}$$

where C is the Chezy coefficient.

If it is assumed that the water infiltrates within the flow boundary layer at near zero velocity in the direction of flow movement, it could be argued that the change of momentum induced by q_I in Eqn. (4) is balanced by an equivalent increase in the bottom shear, resulting in a momentum equation without the term in $q_I(x,t)$. Nevertheless, here that term remains, eliminating it will simplify the analysis.

Eliminating the velocity from the momentum equation, retaining q and y as the dependent variables, differentiating Eqn. (3) with respect to x and Eqn. (4) with respect to t and combining them with Eqn. (5), the following second order partial differential equation of motion results:

$$(gy^3 - q^2)\,\frac{\partial^2 q}{\partial x^2} - 2qy\,\frac{\partial^2 g}{\partial t\,\partial x} - y^2\,\frac{\partial^2 q}{\partial t^2} = \frac{2gq}{C^2}\,\frac{\partial q}{\partial t} \tag{6}$$

$$+ 2q\left[\frac{\partial q}{\partial x}\frac{\partial y}{\partial t} - \frac{\partial y}{\partial x}\frac{\partial q}{\partial t}\right] - (gy^3 - q^2)\,\frac{\partial q_I(x,t)}{\partial x}$$

$$+ 3gy^2(S_o - \frac{\partial y}{\partial x})\,\frac{\partial q}{\partial x} + 3gy^2\,(S_o - \frac{\partial y}{\partial x})\,q_I(x,t)$$

$$- 2yq_I(x,t)\,\frac{\partial q}{\partial t}$$

The above equation is highly non-linear. It will be linearized by assuming that the infiltration depth is small relative to some nominal depth y_o and that discharge and depth are perturbed as:

$$q \equiv q_o + \delta q \qquad\qquad q_o \gg \delta q \tag{7}$$

$$y \equiv y_o + \delta y \qquad\qquad y_o \gg \delta y$$

Substitution of Eqn. (7) into Eqn. (6) will yield:

$$\left(gy_o{}^3 - q_o{}^2\right) \frac{\partial^2 \delta q}{\partial x^2} - 2q_o\, y_o\, \frac{\partial^2 \delta q}{\partial x\, \partial t} - y_o{}^2\, \frac{\partial^2 \delta q}{\partial t^2} - 3gS_o\, y_o{}^2\, \frac{\partial \delta q}{\partial x} \tag{8}$$

$$+ 2qy_o{}^3\, \frac{S_o}{q_o}\, \frac{\partial \delta q}{\partial t} = -\left(gy_o{}^3 - q_o{}^2\right) K\, \frac{\partial \delta q}{\partial x} + 3gS_o\, y_o{}^2\, Kq_o$$

$$+ 3gS_o\, y_o{}^2\, K\,\delta q - 2y_o\, Kq_o\, \frac{\partial \delta q}{\partial t}$$

where C has been assumed constant and equal to the value corresponding to the reference state, i.e.,

$$C = \frac{q_o}{S_o{}^{1/2} y_o{}^{3/2}}$$

and the infiltration losses have been represented (Burkham, 1970a,b) as:

$$q_I(x,t) = Kq^a$$

with a equal to 1, for mathematical tractability reasons. K is defined as the infiltration coefficient.

To obtain Eqn. (8) all terms involving products of linear functions of δy and δq were ignored. Terms involving products of Kq_o (assumed small) and δq were kept but two terms involving products Kq_o and δy were eliminated for convenience. This selective retention of second order terms (under the linearizing assumptions) is a heuristic way of maintaining the dependence of infiltration on variable q even after linearization. It should be pointed out that the assumption that infiltration is proportional to discharge is arbitrary (although documented) and convenient. An assumption of infiltration proportional to depth is possible although the difficulties of obtaining a solution for that case have not be ascertained.

For given initial and boundary conditions, analytical solutions of Eqn. (8) may be obtained. In this paper, the interest is centered in the response of a channel to an instantaneous input along its length for its posterior utilization in the GIUH. First the response of the channel to an input at its upstream end will be found. The boundary condition implied by the definition of a linear instantaneous response is:

$$\delta q(0,t) = \delta(t)$$

where $\delta(t)$ is the delta function.

Before the application of the delta function, the flow is in steady state. It may be expressed as:

$$q(x,t) = q_1\, e^{-Kx} \qquad t \leq 0$$

where q_1 is the flow at $x = 0$. Then, in terms of the linearization scheme, there exists a perturbation about q_o, even in steady state conditions:

$$\delta q(x,t) = q(x,t) - q_o = q_1 e^{-Kx} - q_o \qquad t \leq 0$$

If the reference flow is assumed equal to q_1, the initial conditions to solve Eqn. (8) are:

$$\delta q\,(x\,,0) = q_o\,e^{-Kx} - q_o$$

and

$$\frac{\partial \delta q\,(x\,,t)}{\partial t}\Big|_{t\,=\,0} = 0$$

The procedure for the solution of Eqn. (8) is based on the Laplace transform method. Harley (1967), O'Meara (1969), Dahl (1981), and Kirshen and Bras (1983) among others, have used the Laplace transform method to solve problems of unsteady flow in open channels. A detailed solution of Eqn. (8) is presented by Diaz-Granados, Bras and Valdes (1983). The expression for the net channel response to an instantaneous input at its most upstream point is:

$$h\,(x\,,t) = \exp(-px\,)\delta(t - x\,/c_1) \tag{9}$$

$$+ \exp(-rt\, +\, zx\,)\,(d\,/a\,)^{0.5}\,x\;\frac{I_1[d^{0.5}((t - x\,/c_1)(t - x\,/c_2))^{0.5}/a\,]}{((t - x\,/c_1)(t - x\,/c_2))^{0.5}}\;u\,(t - x\,/c_1)$$

where

$$a = \frac{1}{gy_o\,(1 - F_o^2)^2}$$

$$c_1 = v_o\, +\, (gy_o)^{1/2}$$

$$c_2 = v_o\, -\, (gy_o)^{1/2}$$

$$b = \frac{S_o}{y_o\,v_o}\,\frac{2 + F_o^2}{(1 + F_o^2)^2}\, +\, \frac{K}{v_o}\,\frac{F_o^2}{1 - F_o^2}$$

$$d = \frac{b^2}{4}\, -\, ac$$

$$c = \frac{K^2}{4}\, +\, \frac{3}{2}\,K\,\frac{S_o}{y_o}\,\frac{1}{1 - F_o^2}\, +\, \frac{9}{4}\,\frac{S_o^2}{y_o^2}\,\frac{1}{(1 - F_o^2)^2}$$

$$p = \frac{S_o}{2y_o}\,\frac{2 - F_o}{(1 + F_o)F_o}\, -\, \frac{3}{2}\,KF_o\, +\, \frac{K}{2}$$

$$r = g\,\frac{S_o}{v_o}\, +\, \frac{gS_o\,F_o^2}{2v_o}\, -\, \frac{3}{2}\,Kv_o\,(1 - F_o^2)$$

$$z = \frac{S_o}{2y_o}\, -\, \frac{K}{2}\, +\, \frac{3}{2}\,KF_o^2$$

$$v_o = \frac{q_o}{y_o} = \text{reference velocity}$$

$$F_o = \frac{v_o}{(gy_o)^{1/2}} = \text{reference Froude number}$$

$I_1[\cdot] = $ first order modified Bessel function of the first kind

$u(\cdot) = $ unit step function

This solution is valid for Froude numbers less than 1. For Froude numbers between 1 and 2 the first order modified Bessel function of the first kind, $I_2[\cdot]$, will change to the first order Bessel function of the first kind, $J[\cdot]$, whose solution will contain imaginary terms, implying oscillations in the discharge and water surface.

It is important to note that when $K = 0$ Equation 9 reduces to the same solution obtained by Harley (1967) and used later by Kirshen and Bras (1983).

For a fixed value of x, the area under $h(x,t)$, denoted A_h, is equal to e^{-Kx}. It represents the fraction of the perturbation due to the delta function that reaches point x. By definition of the delta function, $1 - A_h$ is the fraction of it that infiltrates along the interval $[0,x]$; if $K = 0$, $A_h = 1$. In the special case in which $x = L$, in which L is the length of the channel, $h(L,t)$ will be referred to as the upstream inflow IUH and will be denoted as:

$$u(t) \equiv h(L,t) \tag{10}$$

If I is the infiltrated percentage of the flow in a channel of length L, the infiltration coefficient may be expressed as a function of I and L:

$$K = -\frac{\ln(1 - I/100)}{L} \tag{11}$$

As a result, for a given value of the infiltration, the losses will be larger as the length of the channel increases.

The response of the channel to an instantaneous input at its most upstream point $h(x,t)$, can be interpreted as the conditional PDF of the time that a drop entering at the upstream end of the channel spends travelling a given distance x, $f_{T \mid X}(x,t)$. This PDF is a mixed type distribution: a continuous part defined by $h(x,t)$ with an area equal to e^{-Kx} and a discrete part given by a spike at infinity with a value of $1 - e^{-Kx}$. Formally:

$$f_{T \mid X} = \begin{cases} h(x,t) & t \geq 0 \\ P_{T \mid X}(x,t) = 1 - e^{-Kx} & t \geq \infty \end{cases} \tag{12}$$

3. THE LATERAL INFLOW RESPONSE

Recalling the derivation of the geomorphological IUH, the PDF of the travel time of a drop entering anywhere is the channel and travelling to its outlet is required.

For a given channel of length L, the landing spot y of the drop must be between 0, the upstream end, and L, the outlet of the channel. The probability that the drop lands at y is the same for all y within the interval $[0,L]$. Let $x = L - y$ be defined as the distance between the landing spot and the outlet. Therefore, the following PDF of x may be established:

$$f_X(x) = \begin{cases} \dfrac{1}{L}, & 0 < x \leq L \\ 0, & \text{otherwise} \end{cases}$$

The PDF of the travel time of a drop landing anywhere along the length of the channel is given by the unconditional PDF of $f_{T\mid X}(x,t)$, denoted $f_{\tilde{T}}(t)$

$$f_{\tilde{T}}(t) = \int_0^L f_{T\mid X}(x,t)\, f_X(x)\, dx$$

or using Equation 12:

$$f_{\tilde{T}}(r) = \begin{cases} r(t) & t \geq 0 \\ P_{\tilde{T}}(t) = 1 - (1 - e^{-KL})/KL & t = \infty \end{cases} \tag{13}$$

where

$$r(t) = \frac{1}{L} \int_0^L h(x,t)\, dx = g_1(t) + g_2(t) \tag{14}$$

and

$$g_1(t) = \begin{cases} \dfrac{c_1}{L} \exp(-pc_1 t), & t \leq L/c_1 \\ 0, & t > L/c_1 \end{cases} \tag{15}$$

and

$$g_2(t) = \frac{(d/a)^{1/2}\exp(-rt)}{L} \int_0^{L_1} x \, \exp(zx) \, \frac{I_1[d^{1/2}((t-x/c_1)(t-x/c_2))^{1/2}/a]}{[(t-x/c_1)(t-x/c_2)]^{1/2}} \, dx \tag{16}$$

where

$$\begin{cases} L_1 = c_1 t, & t \leq L/c_1 \\ L_1 = L, & t > L/c_1 \end{cases}$$

Since a closed form solution of the above integral does not exist, it must be evaluated numerically. The continuous part involves the travel time of a drop that enters the channel anywhere and reaches the outlet, while the value of the spike at infinity is the probability that a drop landing anywhere infiltrates before the outlet. The area under the continuous part, A_r, is equal to $(1 - e^{-KL})/KL$. A_r represents the fraction of the water that enters along the channel and reaches the outlet.

The term $r(t)$ may be interpreted as the lateral inflow response, i.e., the response of the channel to an instantaneous input at every point along its length. As a result, an individual wave will originate at each point. The total response due to the wave fronts is given by Eqn. (15). This response is zero after $t = L/c_1$ since at this time all the wave fronts travelling at the dynamic velocity $c_1 = v_o + (gy_o)^{1/2}$, have reached the outlet of the channel. The total

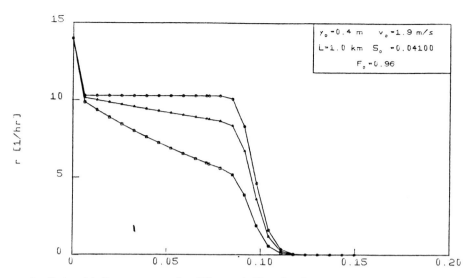

Figure 1. Lateral inflow response for different infiltration losses

    ○   I = 0.0%
    △   I = 10.0%
    □   I = 30.0%

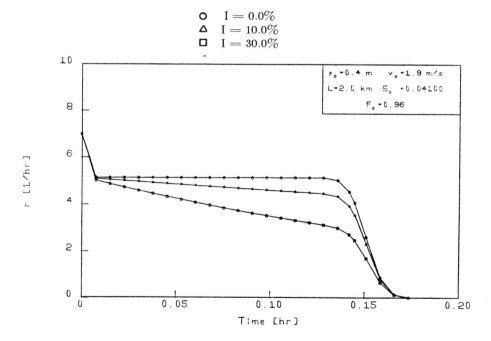

Figure 2. Lateral inflow response for different infiltration losses
    ○ I = 0.0%
    △ I = 10.0%
    □ I = 30.0%

response due to the wave bodies is given by Eqn. (16). In this equation, for $t \leq L/c_1$, the upper integration limit is $L_1 = c_1 t$, which means that waves originating between the outlet and L_1 can contribute to the response at the outlet at time t, however, those waves starting beyond L_1 cannot yet contribute. For $t > L/c_1$, all waves are contributing to the response and the upper limit changes to $L_1 = L$.

Similarly, to the upstream inflow response, if I is the infiltrated percentage of the flow in a channel of length L, the corresponding infiltration coefficient can be expressed as an implicit function of I and L:

$$KL = \frac{1 - e^{-KL}}{1 - I/100} \tag{17}$$

Plots of $r(t)$ for infiltration losses of 0, 10 and 30 percent, and different characteristics of the channel are presented in Figures 1 to 4. As it can be observed, the ordinate of each curve starts at the corresponding value of c_1/L (see Eqn. (14)), independently from the infiltration losses. Figures 1 and 2 correspond to a very steep channel (the reference Froude number is 0.95), for two values of its length, i.e., 1 and 2 km, respectively. In both the response is very fast, and a high percentage of the drops respond before $t = c_1$; for $I = 0$, the shape of the response is basically rectangular; however, as I increases, it tends to decay. Figure 3 shows the responses for a less steep channel with a length of 1 km and reference depth and velocity of 1.0 m and 1.5 m/s., respectively. In this case, they follow closely the shape of an exponential decay. Finally, in Figure 4, the lateral inflow responses are plotted for a longer channel and less rapid reference flow. Their shapes lie between that on Figure 3 and those on Figure 1 and 2.

The fact that the ordinate of $r(t)$ starts at the value c_1/L, independently of I, along with the shape similitude, in some cases, of the lateral inflow response with an exponential decay, allows us to propose a modification to the linear reservoir response (see Eqn. (2)) assumed in the geomorphologic IUH by Rodriguéz-Iturbe and Valdés (1979) and in the geomorphoclimatic IUH by Rodriguéz-Iturbe et al., (1982), in order to take into account the infiltration losses. This is:

$$r_e(t) = \mu e^{-\lambda t} \tag{18}$$

where,

$$\mu = \frac{c_1}{L}$$

and λ is computed such that the area under $r_e(t)$ is $1 - (I/100)$:

$$\lambda = \frac{c_1}{L(1 - I/100)}$$

Figures 5 to 8 present some comparisons between the linearized solution $r(t)$, and the exponential assumption $r_e(t)$, of the lateral inflow response. Figures 5 and 6 show the comparison for a channel with a length of 5k m, a bottom slope of 3 m/km and reference depth and velocity of 1 m and 1.5 m/s, for infiltration losses of 0 and 30 percent, respectively. As it can be seen, the linearized solution responds slower at the beginning, but after approximately 0.2 hours, it becomes faster. In the case of the infiltration losses of 30 percent, the two curves are closer than for zero losses. In Figures 7 and 8, the responses are plotted for a 1 km channel with the same characteristics. The comparison shows similar results, but now the curves are in much closer agreement. In general, the shorter the channel and the bigger the losses, the similarity of $r(t)$ and $r_e(t)$ increases.

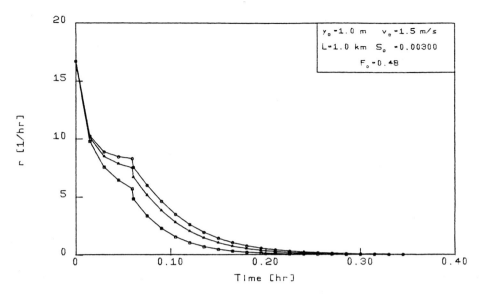

Figure 3. Lateral inflow response for different infiltration losses
○ I = 0.0%
△ I = 10.0%
□ I = 30.0%

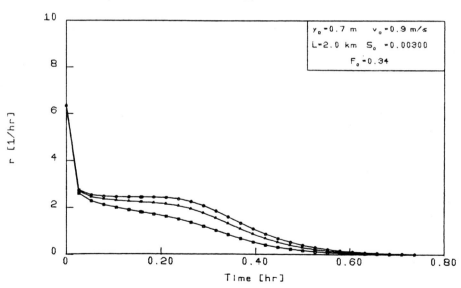

Figure 4. Lateral inflow response for different infiltration losses
○ I = 0.0%
△ I = 10.0%
□ I = 30.0%

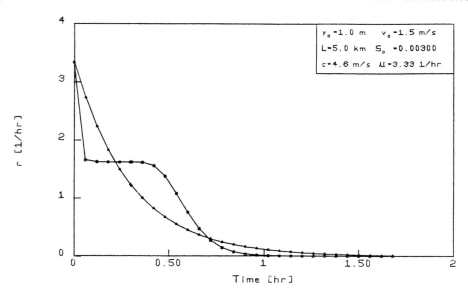

Figure 5. Exponential assumption vs Linearized sclution for the lateral channel response
O Linearized solution
Δ Exponential assumption
I = 0.0%

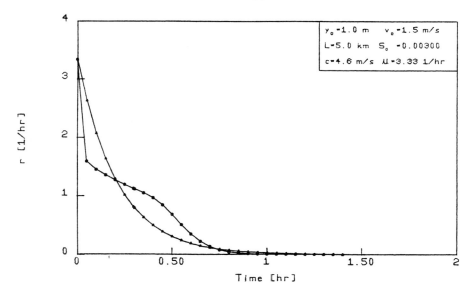

Figure 6. Exponential assumption vs Linearized solution for the lateral channel response
O Linearized solution
Δ Exponential assumption
I = 30.0%

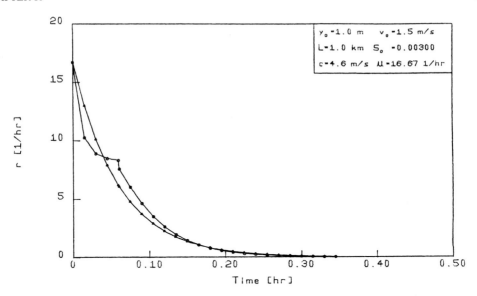

Figure 7. Exponential assumption vs Linearized solution for the lateral channel response
O   Linearized solution
Δ   Exponential assumption
I = 0.0%

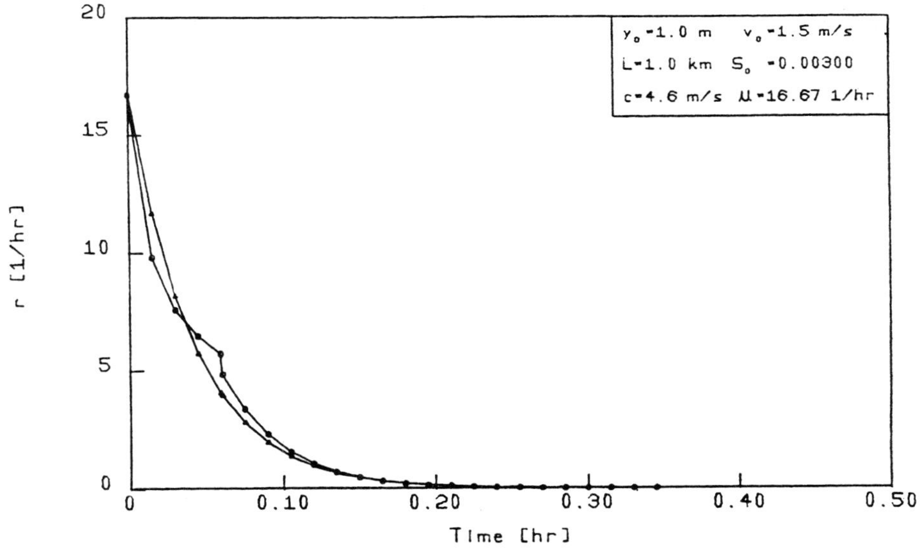

Figure 8. Exponential assumption vs Linearized solution for the lateral channel response
O   Linearized solution
Δ   Exponential assumption
I = 30.0%

It must be noted that although past derivations assume small perturbations, the results will be tested with significant perturbations as is the practice in linear-hydrologic theory. Clearly an error is involved.

4. THE BASIN IUH AND DISCHARGE HYDROGRAPH

Eqn. (13) gives the analytical expression for the PDF of the travel time needed by a drop entering anywhere along the channel to reach the outlet, $f\frac{r}{T}(t)$, which results from the linearized solution of the equations of motion. Replacing $f\frac{r}{T}(t)$ in Eqn. (1) and for a third order basin:

$$h(t) = \Theta_1 P_{12} f\frac{r}{T}^{(1)}(t) * f\frac{r}{T}^{(2)}(t) * f\frac{r}{T}^{(3)}(t)$$

$$+ \Theta_1 P_{13} f\frac{r}{T}^{(1)}(t) * f\frac{r}{T}^{(3)}(t)$$

$$+ \Theta_2 f\frac{r}{T}^{(2)}(t) * f\frac{r}{T}^{(3)}(t) + \Theta_3 f\frac{r}{T}^{(3)}(t)$$

The solution of this equation may be calculated using Laplace transforms. The Laplace transform of $f\frac{r}{T}(t)$ is

$$\left\{ f\frac{r}{T}(t) \right\} = \frac{1}{B_i L_i} \left[e^{B_i L_i} -1 \right]$$

$$B_i = - (a_i s^2 + b_i s + c_1)^{1/2} + e_i s + f_i + K_i$$

In the above expressions, the subscript i indicates the order of the channel, K is the infiltration coefficient, a, b, c have been already defined, and e and f are

$$e = \frac{v_o}{g y_o (1 - F_o^2)}$$

$$f = \frac{K}{2} + \frac{3}{2} \frac{S_o}{y_o} \frac{1}{(1 - F_o^2)}$$

Consequently, the GIUH becomes:

$$h(t) = \Theta_1 P_{12} \; \pounds^{-1} \left\{ \frac{1}{B_1 L_1} \left(e^{B_1 L_1} -1 \right) \frac{1}{B_2 L_2} \left(e^{B_2 L_2} -1 \right) \frac{1}{B_3 L_3} \left(e^{B_3 L_3} -1 \right) \right\} \quad (19)$$

$$+ \Theta_1 P_{13} \pounds^{-1} \left\{ \frac{1}{B_1 L_1} \left(e^{B_1 L_1} -1 \right) \frac{1}{B_3 L_3} \left(e^{B_3 L_3} -1 \right) \right\}$$

$$+ \Theta_2 \; \pounds^{-1} \left\{ \frac{1}{B_2 L_2} \left(e^{B_2 L_2} -1 \right) \frac{1}{B_3 L_3} \left(e^{B_3 L_3} -1 \right) \right\}$$

$$+ \Theta_3 \, \pounds^{-1} \left\{ \frac{1}{B_3 L_3} \left(e^{B_3 L_3} - 1 \right) \right\}$$

Unfortunately, the above equation cannot be solved analytically, but numerically. The same occurs with the discharge hydrograph $Q(t)$, which results from the convolution of $h(t)$, as given by Eqn. (19), with a rainfall event, represented with a net intensity i_e, constant through a duration t_e. The corresponding expression for $Q(t)$ is

$$Q(t) = \frac{\Theta_1 P_{12} A_3}{b} \, \pounds^{-1} \left\{ \frac{1}{B_1 L_1} \left(e^{B_1 L_1} - 1 \right) \frac{1}{B_2 L_2} \left(e^{B_2 L_2} - 1 \right) \frac{1}{B_3 L_3} \left(e^{B_3 L_3} - 1 \right) \frac{1 - e^{-t_e s}}{s} \, i_e \right\}$$

$$+ \frac{\Theta_1 P_{13} A_3}{b} \, \pounds^{-1} \left\{ \frac{1}{B_1 L_1} \left(e^{B_1 L_1} - 1 \right) \frac{1}{B_3 L_3} \left(e^{B_3 L_3} - 1 \right) \frac{1 - e^{-t_e s}}{s} \, i_e \right\}$$

$$+ \frac{\Theta_2 A_3}{b} \, \pounds^{-1} \left\{ \frac{1}{B_2 L_2} \left(e^{B_2 L_2} - 1 \right) \frac{1}{B_3 L_3} \left(e^{B_3 L_3} - 1 \right) \frac{1 - e^{-t_e s}}{s} \, i_e \right\}$$

$$+ \frac{\Theta_3 A_3}{b} \, \pounds^{-1} \left\{ \frac{1}{B_3 L_3} \left(e^{B_3 L_3} - 1 \right) \frac{1 - e^{-t_e s}}{s} \, i_e \right\} \tag{20}$$

where A_3 is the area of the basin, t_e is given in hours, i_e in cm/hr. and b is a conversion factor equal to 0.36 in order to obtain $Q(t)$ in m^3/sec .

5. PARAMETER ESTIMATION

In order to calculate the IUH derived from the linearized solution for a given basin, two sets of parameters must be estimated: parameters representing the physiographic characteristics of the basin and individual channels and parameters representing the dynamic component of the response.

The physiographic characteristics of the basin are expressed in terms of the Horton's numbers, R_A, R_B and R_L. The characteristics of the individual channels are lumped according to the stream order. The average channel length and the geometric mean of the slopes, by stream orders, are used to represent the channel's physiographic characteristics. All the above parameters may be estimated easily from topographic maps, areal photographs, or satellite imagery.

The reference depth y_o, reference velocity v_o and the infiltration factor, K, represent the dynamic component of the response. These parameters, are also lumped according to the order of the stream, and their estimation may involve field inspections and some engineering judgment. In the following lines, a procedure to estimate y_o and v_o is proposed, based on the Manning's equation, and on the expressions $c_1 = v_o + (g y_o)^{1/2}$ and $F_o = v_o / (g y_o)^{1/2}$:

1. From visual inspection, estimate, for each order stream i, the average Manning's roughness coefficient n and the Froude number in steady state conditions, F_o.

2. Using the estimated values of S_o for each order, calculate the respective values of the celerity of the wave flood as

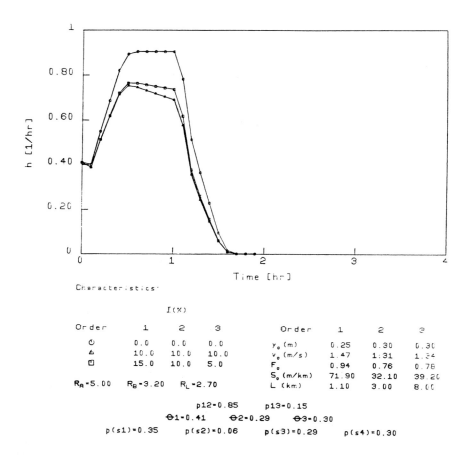

Characteristics:

		$I(\%)$						
Order	1	2	3		Order	1	2	3
○	0.0	0.0	0.0		y_o (m)	0.25	0.30	0.30
△	10.0	10.0	10.0		v_o (m/s)	1.47	1.31	1.34
▢	15.0	10.0	5.0		F_o	0.94	0.76	0.78
					S_o (m/km)	71.90	32.10	39.20
R_A=5.00	R_B=3.20	R_L=2.70		L (km)	1.10	3.00	8.00

$p12$=0.85 $p13$=0.15

$\theta1$=0.41 $\theta2$=0.29 $\theta3$=0.30

$p(s1)$=0.35 $p(s2)$=0.06 $p(s3)$=0.29 $p(s4)$=0.30

Figure 9. Basin iuh for different infiltration losses
(Basin representation 1)

$$c_1 = n^3 F_o^3 g^2 (1 + F_o) / S_o^{1.5} \tag{21}$$

3. Calculate y_o and v_o for each order stream.

$$y_o = \frac{c_1^2}{g(1 + F_o)^2} \tag{22}$$

$$v_o = \frac{c_1 F_o}{1 + F_o} \tag{23}$$

where g is the gravitational acceleration in m/sec ^{2}, and the units of c_1 and v_o are m/sec, and y_o is given in m.

The procedure can be modified slightly, in the case an estimate of the celerity of the wave is available, and therefore in step 2, instead of calculating c_1, Eqn. (21) can be solved for F_o by trial and error.

Finally, the infiltration coefficient K, may be estimated from isolated streamflow measurements performed in reaches where no inflows from tributaries are present. From these measurements, the percentage of the flow infiltrated can be evaluated and introduced into Equation 17 to obtain the estimate of K.

6. HYDROGRAPHS FOR THREE BASINS

The GIUH's and discharge hydrogrpahs for three basins were studied. The first two correspond to sub-basins of the Indio basin, located in Puerto Rico, namely Morovis and Unibon basins, which have been studied in the context of the geomorphologic IUH by Valdés et al. (1979) and Kirshen and Bras (1983). Rodriguéz-Iturbe et al. (1979) present for these basins the main characteristics of several discharge hydrographs determined by a rainfall-runoff model, based on the continuity equation and the kinematic wave approximation to the equation of motion, some of them will be used to check the hydrographs obtained here, by comparing their main characteristics. The third basin is Wadi Umm Salam, also studied by Kirshen and Bras (1983). This is a sub-basin of Wadi A bad, one of the largest wadis in Upper Egypt. Wadis like Wadi Umm Salam are subject to occasional flash floods, which cause damages to the downstream villages. Usually there are no rainfall measurements at any location within the wadi, nor are streamflow measurements. Thus, the geomorphologic IUH constitutes a useful tool to estimate the discharge due to a specific storm in these wadis since only a topographic map or areal photographs, estimation of the storm characteristics, and perhaps a field inspection are required.

Figures 9, 10, and 11 show the GIUHs for the above basins using different combinations of the infiltration losses in the channel. Each figure contains the information about these values of I, and the physiographic characteristics of the basin and the channels. The values of v_o and y_o were estimated according to the modified procedures proposed, assuming a velocity of the wave flood of 3 m/sec and an estimated value of the Manning's coefficient of about 0.067 for Morovis and Unibon (very steep channels, presumably with big rocks in the bed), and about 0.045 for Wadi Umm Salam. The responses of Morovis and Unibon are very similar, the first one being a little faster. The response of Wadi Umm Salam is slower, since it is not as mountainous as the others. The effect of the infiltration losses are clearly illustrated with the differences in the height and area under the IUHs.

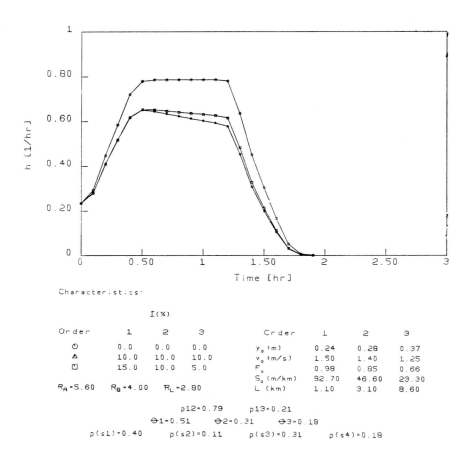

Figure 10. Basin iuh for different infiltration losses
(Basin representation 1)

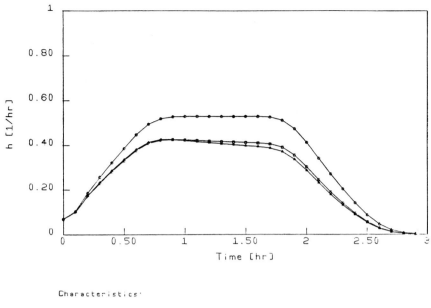

Characteristics:

	I(%)					1	2	3
Order	1	2	3	Order				
○	0.0	0.0	0.0	y_o (m)		0.39	0.40	0.41
△	10.0	10.0	10.0	v_o (m/s)		1.05	1.01	0.98
◻	15.0	10.0	5.0	F_o		0.54	0.51	0.49
				S_o (m/km)		8.00	7.00	6.50
R_A=5.00	R_B=4.00	R_L=2.80		L (km)		1.30	3.60	10.00

p12=0.79   p13=0.21

Θ1=0.64   Θ2=0.30   Θ3=0.06

p(s1)=0.50   p(s2)=0.14   p(s3)=0.30   p(s4)=0.06

Figure 11. Basin iuh for different infiltration losses
(Basin representation 1)

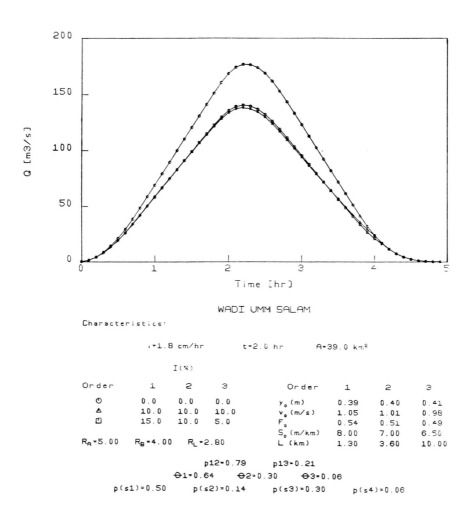

Figure 12. Discharge hydrographs using different infiltration losses
(Basin representation 1)

Table 2 presents the comparisons of the main characteristics of the discharge hydrographs between the linearized solution (Eqn. (20) with $I = 0$ percent) and the rainfall-runoff model (Rodriguéz-Iturbe et al., 1979). As it can be seen, the agreement is good. Figure 12 shows the discharge hydrographs for Wadi Umm Salam for an effective rainfall intensity of 1.8 cm/hr. and a duration of two hours.

TABLE 2. Comparisons Between the Rainfall-Runoff Model and the Linearized Solution

Basin	i_e	t_e	Q_p	Rainfall-Runoff Model T_p	Linearized Solution Q_p	T_p
	(cm/hr)	(hr)	(m^3/s)	(hr)	(m^3/s)	(hr)
Morovis	3	2	103	2.2	103	1.5
	3	3	112	3.0	106	1.5
Unibon	3	2	188	2.0	181	1.6
	3	3	194	3.0	183	1.7

Figures 13 to 16 present some results of the discharge hydrographs when the linearized solution (Eqn. (20)) is compared with that resulting from the modified linear reservoir assumption (see Eqn. (18)) which includes infiltration losses. As it is shown in these figures, both solutions gives similar hydrographs, in terms of shape, peak discharge and time to peak, which permits us to conclude that the linearized solution exponential approximation is valid.

6. CONCLUSIONS

The linearization of the open channel flow equations with infiltration losses shown in this paper is not unique. Other representation of the losses as well as linearization schemes are possible. Also, in theory the results are only valid for small perturbations and limited infiltration. Nevertheless, the goal was to obtain a linear approximation to channel response with losses and this has been done in a form analogous to the accepted linear results of Harley (1967), O'Meara (1968), and Kirshen and Bras (1983). The results can be useful new approximations in traditional hydraulic routing applications.

The motivation of the work, though, was to examine the channel retention time assumptions used in the Geomorphologic IUH by Rodriguéz-Iturbe and Valdés (1979). Interestingly enough, a good approximation to their exponential time distribution is obtained in particular for low Froude number flows, short channels and reasonably high (but not too high) infiltration factor K. Low Froude numbers are the prevalent case in natural streams. The dependence on channel length would imply basin behavior which is scale dependent in a non-proportional way. Nevertheless this conclusion is tainted by the fact that the linearization used is less adequate for long channels since ultimately infiltration would become too large. This requires further study.

For short channels and low Froude numbers, the exponential time distribution does indeed

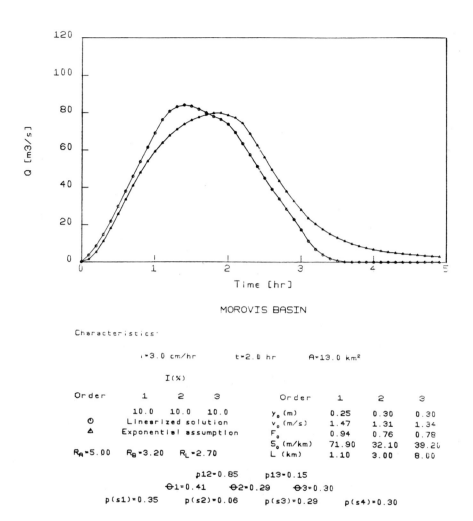

Figure 13. Discharge hydrograph - exponential vs linearized solution
(Basin representation 1)

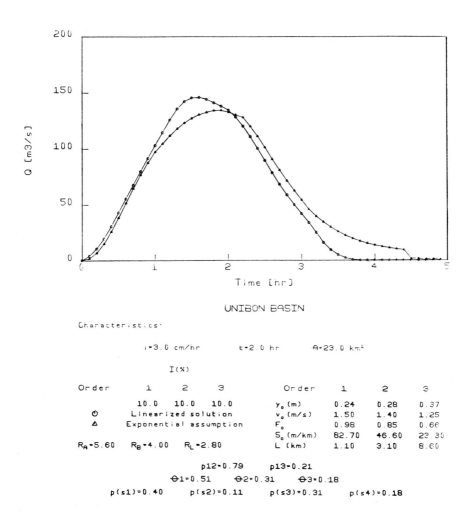

Figure 14. Discharge hydrograph - exponential vs linearized solution
(Basin representaiton 1)

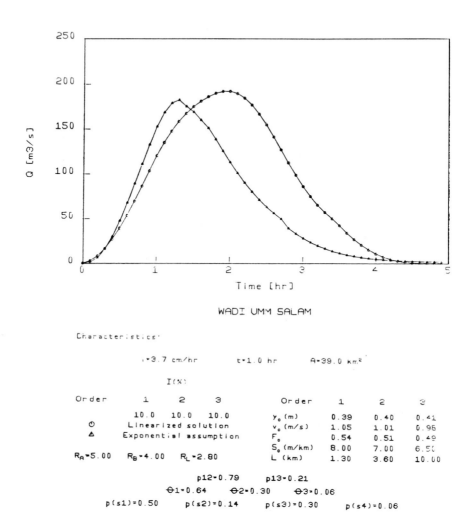

Figure 15. Discharge hydrograph - exponential vs linearized solution
(Basin representation 1)

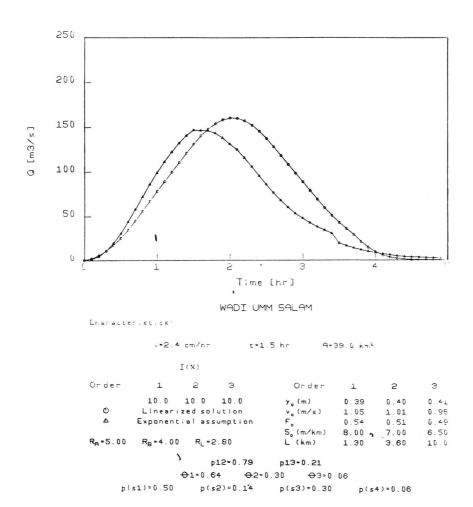

Figure 16. Discharge hydrograph - exponential vs linearized solution
(Basin representation 1)

seem quite adequate. Such linear reservoir analogy is convenient and easy to use. As Eqn. (18) shows, it is simply parameterized by wave celerity and percent of infiltration losses. This parameterization becomes more attractive given the results of Troutman and Karlinger in this issue, where celerity is shown to be the only pertinent dynamic hydraulic parameter.

ACKNOWLEDGMENTS

This study was conducted as part of the MIT/Cairo University Technological Adaptation Program with the support of the Agency for International Development (AID) of the U.S. Department of State. The views and opinions in this paper, however, are those of the authors and do not necessarily reflect those of the sponsors. The authors want to thank Professor Ole Madsen for his useful comments and suggestions.

REFERENCES

Burkham, D.E., 1970a. Depletion of Streamflow by Infiltration in the Main Channels of the Tucson Basin, Southeastern Arizona, *U.S. Geological Survey*, Water-Supply Paper 1939-B.

Burkham, D.E., 1970b. A Method for Relating Infiltration Rates to Streamflow Rates in Perched Streams, Geological Survey Research, *U.S. Geological Survey*, Prof. Paper 700-D.

Dhal, N.J., 1981. Non-steady Flow in Open Channels, Series Paper 10, The Royal Veterinary and Agricultural University, Copenhagen.

Diaz-Granados, M.A., R.L. Bras, and J.B. Valdés, 1983. Incorporation of Channel Losses in the Geomorphologic IUH, Technical Report 293, Ralph M. Parsons Laboratory, Department of Civil Engineering, Massachusetts Institute of Technology.

Harley, B.M., 1967. Linear Routing in Uniform Open Channels, M. Eng. Science Thesis, Department of Civil Engineering, National University of Ireland.

Kirshen, D.M. and R.L. Bras, 1983. The Linear Channel and Its Effect on the Geomorphologic IUH, *Journal of Hydrology*, Vol. 65.

O'Meara, B.E., 1968. Linear Routing of Lateral Inflow in Uniform Open Channels, M. Eng. Science Thesis, Department of Civil Engineering, National University of Ireland.

Pilgrim, P.H., 1977. Isocrones of Travel Time and Distribution of Flood Storage from a Tracer Study on a Small Watershed, *Water Resources Research*, 13(3), pp. 587-595.

Rodríguez-Iturbe, I. and J.B. Valdés, 1979. The Geomorphologic Structure of Hydrologic Response, *Water Resources Research*, 15(5), pp. 1409-1420.

Rodríguez-Iturbe, I., G. Devoto, and J.B. Valdés, 1979. The Geomorphologic Structure of Hydrologic Similarity: The Interrelation Between the Geomorphologic IUH and the Storm Characteristics, *Water Resources Research*, 15(6), pp. 1435-1444.

Rodríguez-Iturbe, I., M. Gonzalez and R.L. Bras, 1982. A Geomorphoclimatic Theory of the Instantaneous Unit Hydrograph, *Water Resources Research*, 18(4), pp. 877-886.

Troutman, B. and M.R. Karlinger, 1986. Averaging Properties of Channel Networks Using Methods in Stochastic Branching Theory, (this issue).

Valdés, J.B., Y. Fiallo, and I. Rodríguez-Iturbe, 1979. Rainfall-Runoff Analysis of the Geomorphologic IUH, *Water Resources Research,* 15(6), pp. 1421-1434.

LIST OF AUTHORS

Keith Beven, Department of Environmental Sciences, University of Lancaster, Lancaster, LA1-4YQ, United Kingdom.

Rafael L. Bras, Department of Civil Engineering, Massachusetts Institute of Technology, Cambridge, MA 02139.

Elpidio Caroni, CNR IRPI, via Vassalli Eandi 18, 10138 Torino, Italy.

C. Corradini, National Research Council, IRPI, Loc. Madonna Alta, 06100 Perugia, Italy.

Mario Diaz-Granados, Department of Civil Engineering, Massachusetts Institute of Technology, Cambridge, MA 02139.

Vijay K. Gupta, Department of Civil Engineering, University of Mississippi, University, Mississippi 38677.

Charles S. Hebson, Department of Engineering Hydrology, University College, Galway, Ireland.

Michael R. Karlinger, U.S. Geological Survey, Denver Federal Center, Box 25046, Mail Stop 420, Lakewood, Colorado 80225.

Mike Kirkby, School of Geography, The University, Leeds, England.

F. Melone, National Research Council, IRPI, Loc. Madonna Alta, 06100 Perugia, Italy.

Oscar J. Mesa, Department of Civil Engineering, University of Mississippi, University, Mississippi 38677.

Edward R. Mifflin, Department of Civil Engineering, University of Mississippi, University, Mississippi 38677.

Ignacio Rodríguez-Iturbe, Graduate Program in Hydrology and Water Resources, Universidad Simon Bolivar, Caracas, Venezuela.

Renzo Rosso, Hydraulic Institute, University of Genoa, via Montallegro 1, 16145 Genova, Italy.

Franco Siccardi, Hydraulic Institute, University of Genoa, via Montallegro 1, 16145 Genova, Italy.

V.P. Singh, Louisiana State University, Department of Civil Engineering, Baton Rouge, LA 70803.

M. Sivapalan, Water Resources Program, Department of Civil Engineering, Princeton University, Princeton, NJ 08544.

Brent M. Troutman, U.S. Geological Survey, Denver Federal Center, Box 25046, Mail Stop 420, Lakewood, CO 80225.

L. Ubertini, National Research Council, IRPI, Loc. Madonna Alta, 06100 Perugia, Italy.

Juan B. Valdés, Graduate Program in Hydrology and Water Resources, Universidad Simon Bolivar, and Instituto de Estudios Avanzados, Caracas, Venezuela.

Ed Waymire, Department of Mathematics, Oregon State University, Corvallis, OR 97331.

Eric F. Wood, Water Resources Program, Department of Civil Engineering, Princeton University, Princeton, NJ 08544.